Modern Stateless Warfare

Also by Paul Brooker

THE FACES OF FRATERNALISM
Nazi Germany, Fascist Italy and Imperial Japan

DEFIANT DICTATORSHIPS
Communist and Middle-Eastern Dictatorships in a Democratic Age

LEADERSHIP IN DEMOCRACY
From Adaptive Response to Entrepreneurial Initiative

NON-DEMOCRATIC REGIMES
Theory, Government and Politics

TWENTIETH-CENTURY DICTATORSHIPS
The Ideological One-Party States

Modern Stateless Warfare

Paul Brooker

palgrave
macmillan

First published 2010
PALGRAVE MACMILLAN

Palgrave Macmillan in the UK is an imprint of Macmillan Publishers Limited, registered in England, company number 785998, of Houndmills, Basingstoke, Hampshire RG21 6XS.

Palgrave Macmillan in the US is a division of St Martin's Press LLC, 175 Fifth Avenue, New York, NY 10010.

Palgrave Macmillan is the global academic imprint of the above companies and has companies and representatives throughout the world.

Palgrave® and Macmillan® are registered trademarks in the United States, the United Kingdom, Europe and other countries

ISBN-13: 978–0–230–21142–1 hardback

This book is printed on paper suitable for recycling and made from fully managed and sustained forest sources. Logging, pulping and manufacturing processes are expected to conform to the environmental regulations of the country of origin.

A catalogue record for this book is available from the British Library.

A catalogue record for this book is available from the Library of Congress.

10 9 8 7 6 5 4 3 2 1
19 18 17 16 15 14 13 12 11 10

Printed and bound in Great Britain by
CPI Antony Rowe, Chippenham and Eastbourne

Contents

List of Tables

Introduction

> 'Political anachronisms can continue to survive through continual modernisation, whether they be non-democratic rule, hereditary monarchy, or even armed rebellion' (Brooker, 2009).
>
> '[The] question only remains whether mankind at large will gain by this further expansion of the element of war; a question to which the answer should be the same as to the question of war itself. We shall leave both to the philosophers' (Clausewitz, 1832).

The two most significant survival stories of modern political history have been the way in which a) non-democratic rule and b) stateless warfare have retained their global prominence through a process of continual modernisation. Both non-democratic rule and stateless warfare are ancient phenomena that were once epitomised by hereditary monarchies and by conflicts between warrior bands. Such ancient and anachronistic phenomena might have been expected to succumb to the modern principles of a) the democratic rule of the people and b) the state's monopoly of legitimate violence within its territory. However, the capacity of political anachronisms to survive through continual modernisation has been amply demonstrated by not only the prevalence of modern dictatorships but also the continuing survival of modern stateless warfare, especially its insurgent variant.

Of course modern states have vigorously opposed the 'private wars' of political goals being pursued by the 'other means' of organised violence. Private wars between the semistates of feudal lords or between wholly stateless bands have therefore either become historically extinct or been modernised but globally marginalised. In their modernised form they are typically Third World civil wars or conflicts between modern warlords, militias, paramilitaries or roaming bands of 'warriors' armed with modern small arms.

On the other hand, the vertical variant of stateless warfare – insurgency by a band against a state – has been modernised but certainly *not* marginalised. The distinctive, vertical nature of insurgent stateless warfare is evident in the linguistic derivation of 'insurgency' from the Latin *insurgere*, 'to rise up'. And modern insurgents have often risen up, sometimes with surprising success, against even very powerful states. Insurgent stateless warfare has therefore remained globally prominent and bears comparison with the way in which non-democratic rule has been modernised and has retained its global prominence. Such evolutionary similarities might well be expected of rule and warfare, though, considering how much they have in common, even being similarly categorised as a) rule by a democracy, semidemocracy or non-democratic regime and b) warfare by a state, semistate or stateless band.

Both non-democratic rule and insurgent stateless warfare began adapting to modernity some two centuries ago and in both cases have gone through three distinct phases of modernisation. In the case of non-democratic regimes it began with Napoleon using referendum-like plebiscites, supposedly expressions of the 'will of the people', to legitimise his non-democratic rule as military dictator of France. The second phase of modernisation added an ideology and a political party to create the distinctively 20th century dictatorship, the ideological one-party state, which politically and militarily rivalled democracy for much of the century. Finally, the third, recent phase has seen the abandonment of ideology and one-partyism in favour of camouflaging dictatorship as a multiparty democracy. After the 1970s–90s global wave of democratisation, it was clear that the long-term future of dictatorship would lie in avoiding open rivalry with democracy and in returning to the first phase's approach of 'if you can't beat them, [appear to] join them' but now with semi-competitive elections giving a much more credible democratic disguise than the old-fashioned plebiscitary referendum or one-candidate election.

The three phases in the modernisation of insurgent stateless warfare are less discernible or clear-cut but are easier to label: 1) the nationalist phase, 2) the maoist-model phase and 3) the new-techniques-and-technology phase. As is described in Chapter 1, each introduced one of the three features of modern warfare identified by Wright in his classic study of war. He pointed out that warfare has been modernised from the 15th century onwards by the introduction of new techniques and new technology, the intensification of military activities and the new emphasis on nationalism and national sovereignty. In the case of stateless warfare, these three features did not begin to appear until the 19th century and would take the form of three separate and sequential phases of modernisation, with the second phase not appearing until the 1930s and the third beginning to emerge in the 1960s. However, the appearance and history of new phases of modernisation has been complicated by the continuing presence of 'species' that are evolutionary hangovers, like reptiles in the age of mammals, from earlier phases or even from the premodern era.

The first phase in the modernisation of stateless warfare introduced the emphasis on nationalism and national sovereignty. The beginning of this phase can therefore be dated to 1809, when the new notion of nationalism was publicly and internationally invoked by Spain's rebel Central Junta to justify its call for the whole Spanish population to form *guerrilla* (small or small-scale war) bands to stage bandit-like attacks against the French army of occupation. Just as Napoleon had modernised his military dictatorship of France by adopting the democratic legitimating principle of 'the will of the people', so the Spanish insurgents were modernising their fight against Napoleon's army by adopting modern warfare's emphasis on nationalism and national sovereignty. Although nationalism had not replaced the premodern, localist and religious basis for their insurgency, the Spanish insurgents had added a new feature that marked the emergence of modern guerrilla warfare.

The second phase in the modernisation of stateless warfare introduced a new feature that was a more narrowly military factor – the increased intensity of military activity. Somewhat paradoxically, the increased military intensity was achieved, or was planned to be achieved, by shifting to another form of warfare. For the second-phase insurgents sought to shift from guerrilla warfare's small-scale 'hit and run' attacks to semistate warfare's large-scale battles and campaigns aimed at taking and holding territory. Such a shift is famously associated with the Chinese communist revolutionary Mao Zedong, who in the 1930s espoused the new insurgent strategy of 1) destroying the local state by eventually shifting to semistate rather than guerrilla warfare and 2) in the meantime preparing for the shift by establishing semistate base areas and creating regular armies capable of engaging in a semistate form of warfare.

The second, 'maoist' phase of modernisation saw an increase in the intensity of political as well as military activity. This increase in political intensity was similar to the second phase in the modernisation of non-democratic rule, which was characterised by the advent of the ideological one-party state. The maoist-model insurgencies tended to be more ideological and politically organised than their first-phase counterparts, and in fact Mao was leading an insurgent equivalent of the communist variety of ideological one-party state. However, such political intensification was only a tendency, not a defining characteristic, and some second-phase insurgencies have lacked a significant ideology and/or party. Similarly, some second-phase insurgent bands have not been communist revolutionaries but instead have been separatists seeking independence from their local state or anti-colonial nationalists seeking to expel a foreign state.

The third phase of modernisation has used new techniques and technology as a means of increasing military intensity without shifting to semistate warfare. Third-phase insurgencies have thereby avoided the increasingly dangerous maoist-model strategy of preparing (by establishing base areas and creating regular armies) for a shift to semistate warfare's large-scale battles and campaigns against the state. Instead they have preferred a strategy of exerting indirect political pressure through the military 'power to hurt' of their new techniques and technology or they have been content with the typically terrorist strategies of propaganda and provocation.

In this sense the third phase has been a military counterpart of the third-phase modernisation of non-democratic rule, which has avoided open political rivalry with the democracies and has camouflaged dictatorship as multiparty democracy. The third-phase insurgencies have not favoured this political approach, and in fact have not shown a preference for any particular political approach or ideology. But their distinctive military approach is strikingly similar to the non-democratic regimes' political approach, with the third-phase insurgencies abandoning semistate warfare's open military rivalry with the state's army and instead relying upon the insurgents' basic military advantage of being a hidden enemy who is 'everywhere and nowhere'.

In the longer term there is likely to be a firmer link or inter-relationship between the historically parallel paths of modernisation of stateless warfare and modernisation of non-democratic rule. As will be described in the Conclusion, the present rate of technological progress in cyber, biological and nanotechnological warfare suggests a) that by the 2030s the third-phase insurgents may be able to rival a nuclear-armed state's apocalyptic military power to hurt and b) that states' political and security responses to this greatly increased insurgent threat may well lead to the spread of a new, Orwellian form of dictatorship – marking a new, fourth phase in the modernisation of non-democratic rule. If the third-phase insurgencies' acquisition of apocalyptic weapons is likewise viewed as a future, fourth phase of modernisation, then it seems that the two historically parallel paths of modernisation may virtually merge in a few decades' time – with the fourth, apocalyptic phase of insurgency's modernisation producing a fourth phase in the modernisation of non-democratic rule.

So it seems only appropriate to examine modern stateless warfare in a way that 'parallels' the way in which modern non-democratic rule has been analysed, explained and even predicted. The second part of this book therefore applies a new framework to insurgency that parallels the existing motive-means-opportunity calculus of intervention/usurpation which has been used since the 1960s to explain and predict the occurrence of modern non-democratic rule. But it would be quite artificial and possibly dangerous to push the parallels too far. An obvious example is the lack of any equivalent in modern non-democratic rule of the vertical and horizontal variants of modern stateless warfare. The nearest equivalent of this separation into two distinct variants would be the distinction between monarchy and dictatorship or between personal rule and rule by an organisation, but both lack a key feature of the vertical-horizontal distinction – the huge difference in significance between the two variants. In primeval and ancient times the horizontal, inter-band variant of stateless warfare was much more significant than the vertical, insurgent variant and in fact the horizontal was the *only* variant to be found in prestate societies. However, the horizontal, inter-band variant of *modern* stateless warfare has been insignificant and overshadowed by warfare between semistates, while the vertical variant is of great past, present and future significance. The study of modern stateless warfare must therefore be focused on the vertical variant – insurgency – and indeed the first part of the book is titled 'Insurgent Stateless Warfare' to highlight its distinctive focus on the vertical variant.

Similarly, the first part of the book has distinctive concerns and does not follow a framework or set of subject-headings that has been applied to modern non-democratic rule. Its chapters provide an analysis of modern insurgency that is concerned with strategies rather than policies and with operational formats rather than methods of controlling state or society. And these chapters recognise such distinctive features of insurgency as the similarities between its operational formats and those of professional criminals, whether rural bandits or urban gangsters, or the similarities between its strategies and those of professional soldiers engaged in wars between states.

Part I begins with a chapter-length survey of the 200-year-old history of insurgency's modernisation, followed by a chapter that explores the key concepts of insurgency. This conceptual chapter delineates the expeller, revolutionary and separatist types of insurgency and then distinguishes between the guerrilla and the terrorist modes of insurgency. It also provides a conceptual and systematic analysis of the difference between insurgent and horizontal stateless warfare and of the difference between stateless and semistate forms of warfare.

Then Chapter 3 shifts the focus to questions of strategy by exploring at some length the two basic strategies of insurgency – preparation and pressure. The other two strategies, the typically terrorist strategies of propaganda and provocation, are discussed in Chapter 4 after its opening historical comparisons with states' terroristic warfare. The perspective on insurgency changes in Chapter 5, which makes comparisons with criminality rather than warfare, but these comparisons also help to describe insurgency's bandit-like and gangster-like operational formats and how they relate to insurgent strategies. Finally, this chapter introduces the forensic motive-means-opportunity framework that can be adapted and applied to insurgents – and is indeed the focus of the four chapters of Part II.

The forensic analytical framework of motive, means and opportunity has already been adapted and applied by political scientists to the study of modern non-democratic rule. As long ago as the 1960s it was used to explain and predict military coups, and was labelled the 'calculus of intervention'. Recently it was adapted and expanded to include not only military coups but also political parties' seizure or misappropriation of power, and was re-labelled the 'calculus of usurpation'. When applied to modern stateless warfare rather than modern non-democratic rule, the forensic framework will therefore be labelled 'the calculus of insurgency' and will be used to explain and predict the occurrence of insurgency in a particular country, region or global era.

The four chapters of Part II focus on the three components or variables of the forensic framework. Chapters 6 and 7 provide an analysis of, respectively, motives and means, while Chapters 8 and 9 describe the various facets of insurgents' opportunity. These two chapters on opportunity incorporate an environmental approach to analysing insurgency that was pioneered by Mao Zedong in the 1930s. Chapter 8 examines not only the rural/urban physical environment but also the insurgents' political and military environments, while Chapter 9 focuses on the social-political environment in both the rural/urban sense and the ethnic/sectarian sense.

Finally, the Conclusion sums up the book's analysis and looks to the future – to the fourth, apocalyptic phase of modernisation. After assessing the apocalyptic means of insurgency that will be available within a few decades, it presents a gloomy prognosis for the middle part of this century. Back in the 1830s Clausewitz asked a crucial question about the rise of modern stateless warfare: 'whether mankind at large will gain by this further expansion of the element of war'. Mao would doubtless have given a positive answer in the 1930s but it seems likely that any military or political theorist of the 2030s will give a decidedly negative answer.

Part I

Insurgent Stateless Warfare

1
Historical Background

The most influential theorist of warfare provided two famous definitions of war in his book *On War* that can readily be applied to stateless warfare. At first Clausewitz described war as analogous to a duel or a pair of wrestlers and defined war as 'thus an act of force to compel our enemy to do our will'.[1] Of course this definition could be applied to a gangland war between bands of gangsters, and without the force aspect could be used to distinguish 'price wars' from normal business competition between rival firms. So it is not surprising that Clausewitz also provided a more precise definition that can be applied only to force used by states or by groups seeking state-like, political goals. For some pages later he famously defined war as 'the continuation of policy by other means' and added that war is 'a true political instrument, a continuation of *political activity* by other means'.[2] These other means are now commonly specified to be *organised violence*.[3] And a contemporary Clausewitzian strategist has pointed out that even when warfare's organised violence is motivated by religious fervour or other powerful non-political motives, its goals retain the political dimension that is characteristic of war.[4]

The Clausewitzian definition of war applies not only to wars between states but also to stateless warfare, which seems to have been with us since the beginning of our species. An evolutionary biologist has suggested that it would 'not be all that far off the mark' to depict ancient hominid life as involving 'small bands of australopithecines roaming the countryside, huddling for protection together at night – and occasionally meeting up with rival bands, with whom they fought for drinking rights at the waterhole'.[5] Such fighting between social groups for water or other collective goods could be termed a 'political' activity, and by Neolithic times it seems to have been devolved to specialised bands of warriors, as cave paintings appear to show a band of warriors armed with bows and a fight between opposing groups of archers.[6] Such warrior bands seem to have fought quite ritualised and restrained battles against each other but Keeley's provocative *War Before Civilization: The Myth of the Peaceful Savage* has contended that prestate warfare also included the ruthless raiding of prestate societies by bands of warriors that showed little restraint in their treatment of enemy non-combatants.[7] He argues that prestate warfare was 'total war conducted with very limited means' and that from the prestate to the modern era: 'War has always been a struggle between peoples, their societies, and their economies, not just warriors, war parties, armies and navies'.[8]

The emergence of the first states during the Bronze Age meant that warfare between bands began to be overshadowed – not only by warfare between states but also by bands' warfare *against states*. Wherever states emerged, inter-band warfare was soon overshadowed by this new, vertical variant of stateless warfare, in which bands of warriors were 'rising up' as insurgents to fight against a local or foreign state. However, there was still a long way to go before the emergence of *modern* insurgent stateless warfare, and in fact it would be thousands of years before modern 'guerrilla' warfare appeared.

Becoming modern 'guerrilla' warfare

The premodern history of insurgency certainly produced many examples of this variant of stateless warfare. Asprey's monumental history of guerrilla warfare points out that the ancient world saw 'countless examples of guerrilla actions, usually of an independent type undertaken in self-defence by nomads and peasant bands, and normally resulting in little more than temporary embarrassment to the incumbent ruler or temporary harassment to the invader'.[9] The most famous ancient-world insurgencies against a ruler or invader are the uprisings of Jews in ancient Palestine against the Seleucid empire and then later the Roman empire.

These Jewish insurgencies were religious rather than merely localist (tribal, clan or village) and they developed into sophisticated revolts – shifting to semistate warfare or operating in urban as well as rural areas. The 'revolt of the devout' against the Seleucid repression of Jewish religious practices began in 167 BCE under the leadership of the Maccabean family and engaged in a three-year 'guerrilla campaign in the Judean mountains' before developing the military strength to capture Jerusalem.[10] The shift to semistate warfare proved disastrous when the rebels' army was defeated in battle by a Seleucid force but the

Maccabean brothers eventually established an independent Jewish state that lasted until defeat by a Roman army in 63 BCE produced nearly 70 years of indirect rule by Rome. When Judea was finally converted from a client state into a province of the Roman Empire, the change in status from indirect to direct rule produced another religious insurgency – the Zealot movement. It espoused a 'highly effective religious ideology of eschatological struggle for liberation' but was fighting a holy war 'which had to be conducted by means of guerrilla warfare, a situation like the beginnings of the Maccabean rebellion'.[11] Although the Zealots eventually succeeded in 66–70 in stirring up a popular rebellion, it was crushed by the Roman army and was less significant in the history of insurgency than the Zealots' innovative urban operations. It is true that they mainly engaged in attacks on rural settlements from inaccessible bases in desert caves – which led to them being described as bandits and brigands by their opponents – but they also carried out operations in the city of Jerusalem, 'where they executed summary justice' and were labelled *sicarii* (dagger-users) 'based on their tactic of assassination'.[12]

These urban operations foreshadowed the gangster-like operational format, described in Chapter 5, that modern insurgents have often used as a supplement or alternative to a rural bandit-like operational format. The insurgencies produced by the first and second phases of modernisation have sometimes used it to supplement their bandit-like rural operational format, while the third-phase insurgencies have tended to use it as their main or even sole operational format. If the Zealot assassins had been engaged in a nationalist or revolutionary rather than religious insurgency, they would certainly have to be considered the pioneers of modern stateless warfare.

The ancient-era Zealots' religious insurgency was the peak of premodern insurgency and indeed during the medieval era in Europe the dominant form of non-state war was the semistate warfare of feudal lords. This feudalism has been described as an early example of warlordism: a 'personalistic rule' by non-ideological leaders of armed men who 'take advantage of the disintegration of central authority to seize control of relatively small slices of territory'.[13] Although the warlord's slice of territory will in principle belong to some weak or failed state, a warlord exercises his own state-like rule over the territory he controls. And this *semi*state territorial control is one of the features that distinguish his semistate wars from the state*less* warfare of warrior bands.

Semistate warfare between feudal lords and against their monarchs was eventually eradicated by Europe's monarchical states. In the case of France, the royal monopoly of force had been declared as doctrine by the later 1620s and 'took another half-century to become effective'.[14] These monarchical states' effective claim to a monopoly of legitimate force within their territory would later be viewed by Weber as the sociologically defining characteristic of a state.[15] However, the monarchical states were capable of enforcing their claim only because they possessed professional armies that could suppress disorder within the state's territory as well as defend the realm against neighbouring monarchs. And with war being monopolised by states' professional armies, Europe had superseded not only the semistate warfare of feudal lords but also the stateless warfare of warrior bands: the 'emergent European dynastic states' had produced 'environments unfavourable either to guerrilla or quasi-guerrilla tactics'.[16] Yet it was near the end of this era and in this region of the world that modern stateless warfare first made its mark, with the appearance of the first phase of its modernisation.

As was described in the Introduction, the modernisation of stateless warfare involved three distinct phases that each introduced one of the three defining features of modern warfare. These features were identified by Wright in his monumental *A Study of War*, which noted that since the 15th century warfare has been modernised by the introduction of new techniques and new technology, an increase in the intensity of military activities and a new emphasis on nationalism and national sovereignty.[17] In the case of stateless warfare, these three features did not begin to appear until the 19th century, and the three phases have followed a different historical sequence from the modernisation of warfare between states.

For it was the first phase that introduced 'the sentiment of nationalism' and the emphasis on national sovereignty as 'the prevailing value, the dominant sentiment, the political objective, and the leading cause of war'.[18] They were introduced into modern stateless warfare when Napoleon's 1808 attempt to take over Spain provoked a nationalist uprising against the French armies that were invading and occupying the country. In rural areas the uprising included guerrilla bands spontaneously sprouting up and attacking French supply convoys, couriers and other targets of opportunity. The uprising's political organ, the Central Junta, quickly exploited this development by publishing in January 1809 a *Manifesto of the Spanish Nation to Europe* that claimed an explicitly nationalist legitimacy for the mobilisation of civilians for such guerrilla warfare against a foreign invader, and in April the Junta actually 'ordered all able-bodied males in occupied territory to form guerrilla bands'.[19]

The new term *guerrilla* (small war or war on a small scale) was being applied to these bands because they seemed to be engaged in a new form of warfare. It involved the raids, ambushes and other 'hit and run' attacks that are now associated with the notion of 'guerrilla warfare', and a member of the French army in Spain later described the guerrillas as being invisible enemies who were not met in battle and dispersed or concentrated according to their leaders' requirements.[20] However, this was actually an ancient, bandit-like form of warfare that was being revived in a more modern form. Like their ancient-world counterparts, the Spanish guerrillas were denigrated by their opponents as being merely bandits or brigands. Even their British ally at first described them as a patriotic form of brigand:

> among the reports in the British press of the continual disasters experienced by the Spanish armies at the hands of their much more powerful and numerous French foes, there were accounts of a covert 'brigand' force that was causing havoc with the French convoys. ... Obviously, there were sizeable bands of Spanish 'patriots' waging a merciless war deep within Spain.[21]

The Spanish nationalists, including the Central Junta, at first described the guerrilla bands as *corso terrestre*, which can be translated as 'land pirates' or 'land privateers'.[22] The term *corso* is better translated as 'privateer' than 'pirate' in this context since privateers possessed a state-issued commission, the Letter of Marque, to prey in pirate-like fashion upon the shipping of an enemy state and were engaged in what now seems a contracted-out or privatised version of naval warfare – indeed the privateer ship has been aptly described as a 'private ship of war'.[23] On land its nearest equivalent would be a private warband commissioned to prey in bandit-like fashion upon the enemy state's lines of supply and communication or, as in the Spanish case, upon the lines of supply and communication of the enemy state's invading armies. As the nationalist uprising had tacitly and then later officially commissioned the guerrilla bands to carry out such bandit-like attacks, the comparison with privateers was very apt and was more descriptive than the term *guerrilla*.

On the other hand, it is only appropriate that guerrilla warfare takes its name from the Spanish example, which was truly a watershed in the history of stateless warfare – the beginning of its modernisation. In that sense the Spanish nationalists were quite correct to view the guerrilla fighter as engaged in a new form of warfare because 'the irregular fighter taking up any weapon at hand to expel invaders' appeared the 'incarnation of the nation at war'.[24] In appearing to be the incarnation of the nation at war, the Spanish guerrillas had added a new and modern element to stateless warfare. Localist and religious factors may have played as significant a role in the Spanish guerrilla war as they had in the premodern insurgencies; there was local opposition to the French intruders and also religious opposition from Catholics opposed to the French army's anticlericalism. But a new, nationalist theme had been added and it was imbued with the new conception of the nation as being both *a* people and *the* people. Thus some 20 years later Clausewitz referred to guerrilla warfare as 'the people in arms' or 'the people's war' and made a prophetic remark about the new phenomenon:

> In the civilized parts of Europe, war by means of popular uprisings is a phenomenon of the nineteenth century. ... the question only remains whether mankind at large will gain by this further expansion of the element of war; a question to which the answer should be the same as to the question of war itself. We shall leave both to the philosophers.[25]

However, Clausewitz's notion of the people's war envisaged a nationalist people fighting alongside their local state in a war against foreign invaders. He noted that 'this sort of warfare' should be thought of 'within the framework of a war conducted by the regular army', which included detaching small units of regulars to support and encourage the insurgents: otherwise 'the local inhabitants will usually lack the confidence and initiative to take to arms'.[26] Even the Spanish guerrillas had operated alongside some of the Spanish state's regular forces as well as alongside an allied British army that had been sent to the region to combat the French. In fact it has been a characteristic tendency of first-phase modern insurgencies to operate alongside their local state's regular forces or its allies' regular forces. (As late as the Second World War there are such examples as the French Resistance insurgents operating alongside Free French and allied Anglo-American regular forces in the fight against German forces stationed in France.) But in the second phase of modernisation insurgencies would be seeking to create their own regular forces and would often be fighting against their local state.

The nationalist phase was only the first of at least three phases of modernisation that have stretched into the 21st century and have greatly complicated the modernisation of insurgent stateless warfare.

The first phase would itself develop such complications as anti-colonial and separatist nationalism that would overlap with the emergence of revolutionary insurgents who foreshadowed the onset of the second, maoist phase of modernisation. Overlapping occurs even between the first phase and the premodern insurgencies thanks to the presence of 'evolutionary hangovers' analogous to the species of reptiles that have continued to survive in our age of mammals or have even produced new species that seem more 'throwbacks' than evolutionary hangovers. Similar evolutionary complications have occurred in the three-phase modernisation of non-democratic rule, which has experienced overlapping between phases and even between premodern and *second*-phase regimes. For example, the second phase of non-democratic rule's modernisation began in the 1920s–30s, with the emergence of the

Table 1.1 Modernising Insurgent Stateless Warfare

Premodern insurgency	=	localist or religious bandit-like war against a state
First-phase modernising	=	*nationalist* (by a people) bandit-like war to expel a foreign state
	a)	anti-invader bands likely allied with their local state's army
	b)	anti-colonial bands fighting to establish a local state
	c)	separatist variant is nationalist bandit-like war to separate from the local state
Second-phase (maoist)	=	*increased intensity of military activities* by shifting to regular, semistate warfare
	a)	strategy of preparation for shift to semistate warfare
	b)	insurgents may be ideological and politically organised revolutionaries
Third-phase modernising	=	*new techniques and technology* used to increase intensity (instead of shifting to semistate warfare to increase intensity)
	a)	strategies aimed at pressure, propaganda or provocation (instead of preparation for shift to semistate warfare)
	b)	tendency towards urban, gangster-like operational format
	c)	insurgents may be revolutionaries, separatists or other type but not very ideological or politically organised

first communist and fascist dictatorships, but the 1920s–30s also saw the creation of the kingdom of Saudi Arabia and its ruling monarchy, which has survived into the 21st century as an evolutionary hangover of premodern tribal/religious monarchy.

However, overlapping is only to be expected of historical phases rather than 'waves' of modernisation. While each phase of a modernisation involves a different phenomenon, historical waves each involve the *same* phenomenon but occur in a different era, such as what Huntington described as the 'three waves' of democratisation that have occurred in modern history – and culminated with the dramatic 'third wave' of democratisation in the 1970s–90s. Unlike waves, historical phases each involve a different phenomenon or feature and although appearing in historical sequence, will see many historical cases of these phenomena occur 'out of sequence' and in the same era as cases embodying another phase's distinctive features. For example, insurgencies with the first phase's nationalist resistance to invaders have occurred in an era when they 'should' be instead or additionally displaying the second, maoist phase's increased intensity of military activities.

The second, maoist phase of modernisation

The introduction of nationalism into stateless warfare was only the first step in an increasing politicisation of insurgency. Asprey noted in his history of guerrilla warfare that its political aspect became 'intrusive' during the 19th century.[27] But another way of describing this politicisation is the tendency to 'ideological construction' that Wright believed was a characteristic of modern warfare:

> In the modern situation far more conceptual construction is necessary to make war appear essential to the survival of anything important. War, therefore, rests in modern civilization upon an elaborate *ideological construction*, maintained through education in a system of language, law, symbols, and values. ... War in the modern period does not grow out of a situation but out of a highly artificial *interpretation* of the situation.[28]

The modern tendency towards ideological construction and interpretation of war meant that by the 20th century the role of nationalism in warfare between states was being supplemented or even supplanted by new ideological themes. For example, the Russian communist regime's attempt to conquer Poland in 1920 was clearly a communist ideological war, and the ideological aspect had been evident in not only the regime leaders' ambitions to spread the communist revolution through Poland to Germany and even Italy but also their use of communist ideological themes to motivate the Russian troops invading Poland and even to win support from sections of Polish society.[29] Some 20 years later communist Russia, under its new title of the Union of Soviet Socialist Republics, would in turn be invaded by Nazi Germany in what was clearly another ideological war – against 'Jewish Bolshevism' and 'the inferior Slav race'. Even democracies have used new ideological themes in describing and justifying their wars. As early as the First World War the US was participating in a war 'to make the world safe for democracy', and in the second half of the century an anti-communist ideological perspective was applied to the Cold War rivalry with the Soviet Union and to such military interventions as the Vietnam war.

Stateless warfare was quicker than inter-state warfare to introduce new ideological themes. In the 19th century the emphasis on nationalism was sometimes directed towards the new goal of establishing a national sovereignty where none had existed before or had existed only in the historical imagination of the insurgents. If the new national sovereignty was sought by a colony fighting for independence from imperial rule, the situation is quite similar to a nationalist guerrilla war against an invading army – in both situations the nationalist insurgents are fighting to expel a foreign state. But the anti-colonial nationalist guerrillas could not hope for any assistance from the regular forces of their local state or its allies, for these nationalists were fighting to *establish* a local state and were also unlikely to receive assistance from foreign states. The US military intervention to assist the Cuban insurgents fighting for independence from Spain was very much the exception to the rule and the resulting 1898 Spanish-American War confirmed that such interventions would involve foreign states in wider and longer-term complications.

Furthermore, the Cuban fight for independence from Spain is also an example of how the goal of national sovereignty could inspire an insurgency to fight and continue fighting *without* assistance from a local or foreign state. The development of a Cuban nationalism and demand for national sovereignty helped the insurgents to withstand a counterinsurgency campaign in 1895–98 that became internationally notorious for its ruthlessness and destructive impact on the rural population. In contrast, the many premodern insurgencies against the 19th century colonial empires had difficulty in becoming and staying united. 'Divided by geography, by rivalries of caste, tribe, clan, or family, their bonds of common culture weak, a unified response based on a shared sense of self-interest, when it could be mustered, seldom survived the first military debacle'.[30] It is true that not all the premodern insurgencies against colonial empires were localist and fragmented; there was also some religious or partially religious resistance that could rely on Islamic beliefs to provide an element of supra-local unity. Whether 'founded on Islamic precepts' or merely given a 'veneer' of unity or a 'glimmer of fleeting unity' by shared Islamic beliefs, these examples of anti-colonial resistance produced some effective guerrilla warfare.[31] For example, their guerrilla attacks helped defeat two 19th century British invasions of Afghanistan and also severely obstructed Russian expansion in the Caucasus during the middle of the century, while effective guerrilla campaigns were waged against France in Algeria in the 1840s and again in West Africa in the 1880s–90s. However, during the 20th century such premodern, religious or localist insurgencies against colonial rule would be almost wholly replaced by the modernisation of stateless warfare. As Porch emphasises, the modernised anti-colonial insurgencies could exploit a new 'national consciousness hitherto conspicuously lacking' in much of the response to colonial imperialism; the 'rising national consciousness reduced the division and apathy which allowed colonial powers to rule'.[32] Guerrillas demanding national sovereignty would now have a much better opportunity to defeat and expel these foreign states.

But the new national sovereignty was sometimes sought by *separatist* rather than anti-colonial guerrillas, and they were therefore fighting against the *local* state rather than against a foreign state. Separatist nationalists were seeking independence from not a colonial empire but from what was considered legally and internationally to be their 'national' state. They were this local state's citizens, not its colonial subjects, and they had the same political rights, such as electing representatives to the national parliament, as any other citizens of that state. For example, the most famous of separatist guerrilla wars, the 1919–21 Irish war of independence from Britain, arose after Sinn Fein nationalists who had won Irish seats in the 1918 elections to the British parliament refused to become Westminster MPs for constituencies in Ireland and instead formed themselves into a separate Irish parliament. Yet the separatist variation of nationalism can still be viewed as a variant of the first-phase modernisation emphasis on national sovereignty and the nationalist desire for an independent state. It is only with the onset of revolutionary rather than nationalist ideologies that the attack on the local state becomes focused on replacing it with a different form of state – a revolutionary state that will help to create a different form of society.

The revolutionary ideologies began to appear in stateless warfare in the 19th century but the early versions were not the basis for a credible, potentially state-threatening insurgency. In the second half of the century some populist and anarchist groups in Russia and Western Europe launched attacks on their local state that were intended to lead to its destruction and replacement by a different form of state or in some cases by the absence of any form of state. But these insurgents were pursuing the typically terrorist strategies of propaganda or provocation, through attacks targeting the head of state or prominent state officials, and were apparently hoping that this propaganda and provocation (of politically inept repression) would lead on to a mass uprising that would destroy the local state and somehow lead on to the attainment of the insurgents' ultimate political and social goals.

Although the communist revolutionaries did not appear until the 20th century, they posed a much more credible threat of destroying the state and replacing it with a revolutionary state. However, the communists' first seizure of power was not achieved by insurgent stateless warfare but instead by a revolutionary takeover of the government and then a revolutionary civil war to defend the new, communist state and enforce its rule upon the whole country. In fact this revolution in Russia involved 1) the 1917 insurrectionary coup in the capital city and other major urban areas, 2) the victorious 1918–21 civil war that the communists won by creating a regular army to defeat the counter-revolutionary armies and conquer the vast rural/regional hinterland of the country and 3) the creation of a one-party state that would help the communists attain their ideological social goals in the 1920s–30s. When communists in other parts of the world sought to stage a revolutionary seizure of power by guerrilla warfare, they would adopt a very different and almost opposite approach – creating a regular army in the rural hinterland in order to conquer the major urban areas and destroy the local state's army on the battlefield.

The Chinese communist leader Mao Zedong's successful use of this approach in the 1930s–40s produced the 'maoist model' of guerrilla warfare and also the second, 'maoist' phase in the modernisation of stateless warfare. Somewhat paradoxically, although Mao is the most famous theorist of guerrilla warfare, he was actually more concerned with the problem of shifting to a more regular form of warfare. He believed that a rural guerrilla war should be focused on creating a regular army which could defeat the state's army in a revolutionary civil war. In his 1936 'Problems of Strategy in China's Revolutionary War' and later writings Mao envisaged the creation of regular armies through local recruitment and by converting guerrilla fighters into regular soldiers:

> To raise the quality of the guerrilla units it is imperative to raise their political and organizational level and improve their equipment, military technique, tactics and discipline, so that they gradually pattern themselves on the regular forces and shed their guerrilla ways. Politically, it is imperative to get both the commanders and the fighters to realize the necessity of raising the guerrilla units to the level of the regular forces.[33]

In turn, the creation of regular armies would require the establishment of large and stable 'base areas' in the rural hinterland to produce recruits and provide them with 'equipment, military technique, tactics and discipline'. Creating these base areas would involve establishing communist political power over a significant area of territory and mobilising the people living within the area.[34]

In other words, the communists would seek to establish semistate areas that would be the revolutionary equivalent of the semistates set up by counterrevolutionaries in Russia during the 1918–21 civil war. These counterrevolutionary semistates were later included in Trotsky's famous theory of revolution, which depicted social revolutions as involving the growth of a dual power/sovereignty that sees both sides or 'powers' – the revolutionary and the counterrevolutionary – controlling a particular section of society. He noted that this revolutionary situation might develop into a large-scale civil war that 'gives to this double sovereignty its most visible, because territorial, expression. Each of the powers, having created its own fortified drill ground, fights for the possession of the rest of the territory'.[35] In the case of Mao's insurgency, the revolutionaries were indeed seeking a territorial expression of sovereignty – rural base areas – that would constitute a 'drill ground' for preparing the regular army that would fight 'for the possession of the rest of the territory' of China.

The other, less appropriate comparison is to view the communist base areas as the revolutionary equivalent of China's warlord semistates, which had been established by the provincial commanders of Western-style regular forces. These internationally famous 'Chinese warlords' had been the de facto rulers of China in 1916–28 and by the mid-1930s still controlled, however weakly, huge areas of the country that were only nominally ruled by the Nationalist central government of China. (In fact the communists' creation of large and stable base areas was only feasible, even in the most inaccessible regions of this vast country, because the central government had direct control over little more than the central-coastal region.) However, the communists' base areas differed fundamentally from the warlord semistates in being a future rival of the Nationalist central government for rule over the whole of China, not just for control of a provincial semistate. Having each created their own drill ground, the communist and Nationalist powers were preparing to fight for the possession of the rest of China's territory.

In the 1946–9 civil war the communists' regular armies would win that fight for the possession of China's territory, forcing the remnants of the Nationalist army and party-state apparatus to seek refuge on the island of Taiwan. By the end of the civil war the communist People's Liberation Army not only numbered millions of regular soldiers but was also equipped with thousands of artillery pieces and hundreds of tanks. From this perspective Mao's guerrilla strategy of the 1930s seems best described as *preparation* for the civil war of the 1940s and, more broadly, as preparation for shifting *away* from guerrilla warfare and towards a regular form of warfare that was the semistate equivalent of conventional inter-state land war – which Wright depicted as primarily oriented towards 'the occupation of enemy territory. Capture or destruction of his armies and fortifications is a means to this end'.[36]

What is more, the shift to semistate warfare involved a dramatic increase in that intensity of military activity which according to Wright is another of the defining features of modern warfare. His conception of 'intensity' included both 1) 'units of military activity occupying as limited a time and space as possible' and 2) the 'loss of life and wealth' arising from this military activity.[37] Although he envisaged the increase in intensity as applying to the military activities of regular armies, it can also be applied to the shift from guerrilla to regular military activities and in fact this produces a much more dramatic increase in intensity. Instead of carrying out a host of small-scale bandit-like raids, ambushes and other 'hit and run' attacks spread over a wide territory and/or a long period of time, the insurgents' forces are now concentrated in an army host that fights only a few large-scale battles in a relatively short campaign – and of course with an accompanying intensification in the loss of life and wealth.

Such intensifications of military activity may appear to be nothing new in the history of insurgent stateless warfare. After all, there had been *pre*modern shifts from stateless to semistate warfare, as when the Maccabean and Zealot insurgents shifted from bandit-like attacks to large-scale battles against the Seleucid and Roman armies. Amongst the first-phase insurgencies, too, there were at least ambitions or visions of shifting to semistate warfare. In the famous case of the Irish separatists of 1919–21, the Irish Republican Army's central headquarters was 'preparing for some sort of transition towards open warfare', and the fact that 'such a transition was [still not] remotely within the IRA's power in 1921' was one reason why its Chief of Staff and Director of Intelligence, respectively Mulcahy and Collins, were pessimistic enough to support a compromise political settlement with the British government.[38]

But Mao's version of intensification was a watershed in both the theory and practice of shifting to semistate warfare. In addition to his historic victory of 1946–9, more than a decade earlier he had published an explicit strategy of preparing for the shift to semistate warfare by establishing base areas and creating a regular army in these politically mobilised semistate areas. By focusing guerrilla warfare on this 'preparation' strategy he had invented the maoist model of insurgency and the maoist-model strategy of guerrilla warfare.

Thus the second phase in the modernisation of stateless warfare is appropriately labelled the 'maoist' phase, which emphasises that its defining characteristic is the focus on increasing the intensity of military activity through the shift to semistate warfare. The change from the first-phase, nationalist modernisation is highlighted by comparing Mao's 1930s conception of guerrilla warfare with Clausewitz's 1830s description of a nationalist people's war in which the local inhabitants usually lack the confidence and initiative to form guerrilla bands unless detachments of their state's regular army are present to encourage and support them. In contrast, Mao is intending to transform the local inhabitants into not just guerrilla bands but a regular People's Liberation Army that will defeat their state's army on the battlefield. And Mao also has a communist ideology and political party with which to mobilise the people in his base areas – to provide these local inhabitants with more initiative and confidence than the regular soldiers of Clausewitz's era could hope to provide.

In that sense the maoist phase of modernisation involved an increase in the intensity of political as well as military activities. If the political aspect of guerrilla warfare had become more 'intrusive' in the 19th century, it became more 'intensive' in the 20th century in terms of the degree of political organisation as well as the emphasis on ideological interpretation. The intensification was linked to the second phase in the modernisation of non-democratic rule, which was characterised by the advent of the ideological one-party states.[39] Such dictatorships possessed not only an official ideology but also an official party that was given a monopoly as literally or effectively the sole political party in a country that was therefore literally or effectively a one-party state. And the model for this phase of modernisation had emerged in Russia during the 1917–21 period of revolution and civil war, with the development of an official Marxist-Leninist ideology and of a party-state dictatorship by a Leninist communist party. Although the fascists would be the first to establish another example – rightist rather than leftist – of ideological one-party state, the communist version would be adopted by many leftists in China and elsewhere as a model and focus for their revolutionary ambitions of replacing the local state. It would also provide the Chinese communists with a model for ruling and mobilising their semistate base areas, which would become ideological one-party semistates. In addition to having an ideologically legitimated claim to rule over their base areas, the communists had a political party that was much more capable than any army of ruling and mobilising the local inhabitants of their base areas.

So it appears that the maoist phase involved a political as well as military modernisation, namely the use of a political ideology and party. But unlike the military intensification, this political intensification was not a defining feature of the second phase. At most it was a characteristic tendency that accompanied the second phase and highlighted important anomalies in the way that the second phase has evolved. For example, the only second-phase case that comes close to being as famous as Mao's success in China is Castro's success in Cuba, but the 1956–9 Cuban insurgency lacked both a proper ideology and a proper party. The nearest thing to an ideology was a pro-democratic determination to overthrow the Batista military regime, and Castro's only political organisation, the Movement of 26 July, seems to have been little more than an auxiliary urban underground movement. It was only after coming to power that Castro and his lieutenant Che Guevara began to espouse Marxism-Leninism and to develop a Leninist party.

Another anomaly is that the maoist model has been adopted by non-communists and even non-revolutionaries. It acquired such global prestige after the victorious 1946–9 civil war that it would be used by not only communists engaged in anti-colonial rather than revolutionary insurgencies but also by *non*-communists engaged in anti-colonial or separatist insurgencies. For example, the FLN would use it against French colonial rule in Algeria in the mid-1950s and the EPLF would use it to separate Eritrea from Ethiopia in the 1980s–90s. But perhaps a bigger anomaly than this non-revolutionary use of the maoist model is that the next phase in the modernisation of stateless warfare would actually abandon the maoist model and use other means to increase insurgencies' military and political intensity.

The third-phase modernisation of insurgent stateless warfare

The third, ongoing phase in the modernisation of insurgent stateless warfare has seen the introduction of Wright's third defining feature of modern warfare – the use of 'new techniques' and 'new technology' – as an alternative to the maoist model's approach to increasing military intensity.[40] Through these new techniques and technology the third-phase insurgents have been able to increase the intensity of their military activity without adopting the maoist model's approach of shifting to semistate warfare – and therefore without adopting the maoist model's strategy of preparing for such a shift. The third-phase insurgents have instead pursued the pressure strategy of exerting indirect political pressure on the enemy state and/or the typically terrorist strategies of propaganda and provocation.

The third phase's approach is similar to a newly prominent feature of warfare between states that Schelling identified in the 1960s – the decade in which the third phase began to emerge. As the use of military force can not only 'weaken an enemy militarily' but also 'cause an enemy plain suffering', governments can engage in a 'coercive diplomacy based on the power to hurt'.[41] And indeed 'we are in an era in which the power to hurt – to inflict pain and shock and privation on a country, not just its military forces – is commensurate with the power to take and hold' territory.[42] The military power to take and hold territory is what the maoist model hopes to achieve through its shift to semistate warfare, while the newly prominent military power to hurt is what the third-phase insurgencies hope to achieve through their use of new techniques and technology. Schelling noted that the power to hurt inheres in 'some of the least impressive military capabilities', including 'the terrorist's bullet'.[43] But beyond the 1960s the terrorist insurgent would seldom be using a mere bullet; the military capability of the terrorist and the guerrilla would be increased by new techniques and technology that greatly increase their power to hurt, as epitomised by the 9/11 attacks' use of airliners as guided missiles against skyscrapers.

Insurgents' new capability meant not only a greater military power to hurt but also a greatly increased intensity of military activity in terms of loss of life and wealth. The new technology available to them in the 21st century would also raise the prospect of something that Schelling had imagined when he speculated about how the prospect of thermonuclear war would appear if a state's nuclear weapons were used very slowly, such as one per day; then the prospect of thermonuclear war 'would look very different, something like the most terroristic guerrilla warfare on a massive scale'.[44] And his simile may well become a reality if new technology gives insurgents an apocalyptic power to hurt and therefore, as is discussed in the Conclusion, creates a fourth-phase strategic situation.

Table 1.2 Analysing Insurgent Stateless Warfare

Phase of modernisation:	first, second, third and fourth (Chapter 1, Conclusion)
Type of insurgency:	revolutionary, separatist or expeller (Chapter 2)
Mode of insurgency:	guerrilla or terrorist (Chapter 2)
Strategy:	preparation, pressure, propaganda or provocation (Chapters 3–4)
Operational format:	bandit-like or gangster-like (Chapter 5)

The third phase's use of new techniques and technology has been associated with not only a change in strategic approach but also a change in operational format, for there has been a move away from guerrillas' traditional bandit-like format and towards an urban, gangster-like operational format. The change was evident in the structural aspect of the new techniques, which has included international, networked and micro-size structures similar to those used by gangsters. And the change was also evident in the tactical aspect of the new techniques, notably the gangster-like use of the car bomb, the roadside bomb or any other example of what is now termed an Improvised Explosive Device. When given some new touches, notably adding a suicide driver to the car bomb, these new tactical techniques could have a devastating power to hurt. For example, in 1983 hundreds of US Marines deployed as peacekeepers in Lebanon were killed by a suicide-car-bomb attack on the apartment block they were using as a barracks. And in 1993 an IRA car bomb in the City financial district of London showed that even an 'old-fashioned' gangster-like car bomb could inflict very expensive damage when parked in the right place.

Of course the urban, gangster-like operational format is nothing new in insurgency. It was used some 2,000 years ago by the Zealot rural insurgency to assassinate enemies in Jerusalem and it has been used in similarly supplementary fashion by some famous examples of first-phase and second-phase insurgency, such as the Irish separatists' IRA operations in Dublin in 1919–21 or the Algerian FLN's terrorist bombings in the city of Algiers in 1956. What is more, it has been the primary rather than supplementary operational format of the significant number of urban insurgencies that occured in the first and second phases, such as the 19th century populist and anarchist insurgencies or the 20th century Tupamaros' revolutionary insurgency in Uruguay – the most successful of the 1960s–70s 'urban guerrilla' insurgencies. Nonetheless, until the third phase of modernisation arrived, insurgency was typically rural rather than urban and was typically bandit-like rather than gangster-like.

In contrast, the third-phase insurgencies are typically urban and gangster-like. Although a significant number of this phase's insurgencies are rural, they are more the exception than the rule – the tendency is towards urban insurgencies. The rural exceptions also face the challenge of adapting urban techniques and technologies to a rural environment and bandit-like operational format.[45] Using a roadside bomb on a rural road may not be too difficult, but communicating through the Internet from remote villages or mountain caves may be more problematic.

Thus the change in operational format is not as stark as the change in strategy – which involves *all* the third-phase insurgencies pursuing a non-maoist strategy. By definition any insurgency pursuing the maoist preparation strategy is a second-phase insurgency, even if it is using typically third-phase 'new techniques and technology'. For these techniques and technology would be linked to the maoist approach of shifting to semistate warfare and not to an alternative approach and strategy that is content with increasing the intensity of military activities solely through new techniques and technology and not also through a shift to semistate warfare. The alternative approach has three different strategies – pressure, propaganda and provocation – that may be used separately or in combinations and they share the common characteristic of not being maoist and, more positively, of being less dangerous and risky than the maoist strategy.

In fact the main reason that the emergence of the third phase can be traced back to the 1960s is that by then the dangers and risks of the maoist strategy were becoming too obvious for potential insurgents to ignore. At this stage the need for a different strategy was more of a factor than was the increased 'power to hurt' of new techniques and technology. The shift to semistate warfare has always risked playing into the enemy army's hands by solving its key counterinsurgency problem of 'persuading or coercing an irregular enemy to come out and fight so that he could be duly slaughtered in satisfactorily large numbers'.[46] And with states' armies being modernised with new techniques and technology, such as helicopter gunships, any attempt by insurgents to rival these armies on the battlefield, by attacking them with semistate regular forces, had become even more likely to result in the insurgents being slaughtered in large numbers. Similarly, preparing for the shift to semistate warfare had become very dangerous; base areas had become too detectable by air reconnaissance and too vulnerable to attack by helicopter-transported troops, even if the base area was established in remote or inaccessible parts of a country.

The increased vulnerability of second-phase insurgencies was symbolised by the 1967 failure and martyrdom of Che Guevara. He was attempting to set up a base area in an apparently remote and inaccessible region of Bolivia when the Bolivian army hunted him down and eliminated his band. The contrast with Guevara's 1956–9 success as a guerrilla commander in Cuba highlighted the increased vulnerability of the maoist-model insurgency – and highlighted the need for a change of strategy to cope with the more modern environments in which insurgents now had to operate. Changing to an urban, gangster-like operational format was not sufficient if the advantages of hiding in the anonymous urban crowd were nullified by an attempt to establish some sort of urban semistate, as would be confirmed in 1972 by the Uruguayan military's destruction of the Tupamaros' insurgency. And in fact whenever insurgents have sought to establish urban base areas or liberated zones, it has eventually proved suicidal and more dangerous than establishing inaccessible rural base areas.

The only solution to the problem was to adopt strategies that allowed insurgents to exploit their traditional advantage of being a hidden enemy who is 'everywhere and nowhere'. This is what the pressure, propaganda and provocation strategies allowed an insurgency to do; they were potentially stealthy strategies that were especially well-suited to camouflaging and disguising the insurgent band in an urban environment incorporating perhaps millions of other human beings.

There was an element of 'back to the future' in adopting these strategies. The pressure strategy, for example, had been used implicitly or tacitly in earlier phases of modernisation and even by premodern insurgencies. Most notably, it was used implicitly or tacitly to pressure the British government into the Anglo-Irish Treaty of 1921 that gave independence to much of Ireland as the Irish Free State, which is one of the few times in history that separatist insurgents have achieved their goal of separating from the local state and establishing their own state. However, it was acknowledged in the previous section that even the maoist strategy had its premodern forerunners – 'there is nothing new under the sun' perhaps in strategy – and that it is the modernising context and explicit presentation of the strategy that makes it 'new'.

What appears to be the first explicit presentation of the pressure strategy was actually made two years or more *after* Mao had presented his revolutionary guerrilla strategy of preparing for semistate warfare. The pressure strategy was presented in 1938, by Mao himself, when writing about fighting a 'protracted war' against the Japanese armies that were invading China (as is discussed in more detail in Chapter 3). Although his conception of protracted war was focused upon the standard maoist strategy of shifting from guerrilla to regular warfare, he also suggested that the increasing war weariness among the Japanese people might force their government to withdraw its armies from China. In other words, the Japanese state would be forced to withdraw its armies because of the indirect political pressure exerted by i) the Chinese people through ii) fighting a protracted war that iii) had induced war weariness in the Japanese people and thus iv) produced political pressure on their government to end this war by withdrawing its armies from China. Mao presumably meant, too, that by fighting a protracted war the Chinese people would be either consciously or inadvertently using a pressure strategy that might prove to be more effective or speedy than the strategy that he was advocating as the main focus of their military efforts, namely preparing for eventually defeating the Japanese armies through the use of regular armies fighting mobile and positional battles and campaigns. In the event, the Japanese would be defeated by neither of these strategies but by the air, sea and land campaigns of the US and Russia (Soviet Union). Nonetheless, later maoist-model insurgencies would defeat foreign states through using – in supplementary or inadvertent fashion – the pressure strategy that Mao had envisioned. The Vietnamese communists supplemented their maoist-model approach with a pressure strategy that in 1954 forced the French state to withdraw its army from its former Vietnamese colony, leaving the northern half to become communist North Vietnam. Then French colonial rule over Algeria came under attack from a maoist-model insurgent band, the nationalist FLN, which inadvertently used a pressure strategy that eventually produced a French withdrawal from Algeria.

By the end of the 1960s the pressure strategy had clearly won another victory, its most dramatic and internationally noteworthy, as the US superpower had begun withdrawing its army from South Vietnam. This successful use of the pressure strategy against a superpower had come only a year or two after the preparation strategy's great symbolic failure – the defeat and death of Che Guevara. So it seems only appropriate to view the late 1960s as the beginning of the third phase of modernisation, in which the intensification of military activity would no longer involve a maoist approach and preparation strategy but instead the use of new techniques and technology in pursuit of a pressure strategy or the typically terrorist strategies of propaganda or provocation. The emergence of the third phase is therefore not linked to a particular insurgency, like the first phase and the Spanish guerrillas, nor linked to a particular theory of insurgency, like the second phase and Mao's theory, but instead is linked to a watershed in the history of insurgent strategy, namely the decade in which the maoist preparation strategy clearly lost its preeminence. By then some of the new techniques and technology were no longer a novelty, such as the car bombs that had been pioneered by Prohibition-era gangsters or the airliner-hijacking pioneered by maoist-model Venezuelan insurgents earlier in the 1960s. But now they would be employed by non-maoist insurgents and they would eventually be joined by truly novel techniques and technologies, such as having a suicide attacker drive the car bomb or fly the hijacked airliner.

Finally, the third phase also adopted new *political* techniques and technologies, such as television and the Internet, instead of following the maoist model's tendency to use a political ideology and party to increase political intensity. In fact they are seeking a different sort of political intensity, not an intrusive politicisation of the insurgency but an intensity of commitment in the individual insurgents that ensures they are willing to kill and be killed. While the third-phase insurgencies retain the modern characteristic of interpreting their war in ideological terms, they tend to have little interest in ideological 'world views' and 'ways of life' like the communists' Marxism-Leninism. Similarly, they tend to have little interest in political parties, especially the tightly organised, Leninist type of party. Instead their political message and level of political organisation tends to 'fit' their particular historical and cultural environment, enabling them to exploit existing dispositions and structures instead of having to create them largely 'from scratch'.

The third phase therefore seems a politically very flexible phase, not only incorporating all types of insurgency but also displaying a varied range of ideological perspectives. (This flexibility seldom extends to adopting the democratic disguise that is typical of third-phase non-democratic regimes; third-phase insurgencies favour disguise and camouflage only in their military approach and strategies, not in their political activities.) At one extreme there are such anti-modern ideological perspectives as the 1980s–90s Unabomber's manifesto against modern technology's destruction of human freedom. At the other extreme there are insurgencies ideologically building upon localist or religious dispositions, such as Islamic religious beliefs. And they may also be organisationally building upon religious or localist social structures, such as tribal and clan structures. These premodern dispositions and structures can be built upon and 'mobilised' through such new political techniques and technology as Internet web-sites – just as new military techniques and technology can intensify the power to hurt.

The third phase has already produced two insurgencies with a dramatically intensified power to hurt that have built their ideological interpretation upon a religious disposition. In fact the often-cited case of the Aum Shinrikyo cult's sarin-gas attack on the Tokyo Underground in 1995 seems at first glance to have been carried out by a religious band pursuing religious goals. But this poison-gas attack sought 'to produce chaos' that would enable the cult 'to seize power in the resulting confusion', and the cult itself was 'a religious/ideological organization: it had a clear anti-American, anti-Western and anti-Semitic ideology'.[47] Al Qaeda is another, much more widely known case of an apparently religious insurgency with a dramatically intensified power to hurt. And it, too, on closer examination reveals an ideological aspect that has been built upon a religious disposition. Osama bin Laden's doctrinal pronouncements clearly contain an ideological interpretation of his war against what is presented as a political as well as religious foe. For example, in his 1998 declaration of war (jihad) on Americans and their allies by the 'World Islamic Front' he refers to Americans as 'the soldiers of Satan' but also politically analyses the occupation of the Arabian peninsula by America and alludes to his anti-Crusader ideology, such as mentioning 'the Judeo-Crusader alliance', 'the brutal Crusader occupation of the Peninsula' and 'these Crusader hordes that have spread through it like locusts'.[48]

Of course Aum Shinrikyo and Al Qaeda are usually described as terrorist rather than insurgent groups, let alone as any form of guerrilla band. And the next chapter will deal with the problematic issue of distinguishing between the guerrilla and terrorist modes of insurgency. It will also distinguish between the various types of insurgency, such as the revolutionary and the separatist, in a more conceptual and systematic manner than in this chapter's survey of the historical background to modern stateless warfare and its three phases of modernisation.

2
The Types and Modes of Insurgency

The Clausewitzian definition of stateless and other forms of warfare as 'the pursuit of political goals by means of organised violence' begs two obvious clarifying questions – *what are* the goals and *how* is organised violence being used to achieve them? In the case of insurgent stateless warfare the answer to the question 'what are the goals' also answers the key analytical question of what are the types of insurgency. The answer to the 'how' question is more complicated because it involves three different subsidiary questions. The first is the key question in all forms of warfare – how do they plan to win? – and refers to the strategy that is being used. The second question is peculiar to stateless warfare and asks about the operational format that is being used, specifically whether it is bandit-like or gangster-like. The third question is relevant to all forms of warfare and asks whether or not the organised violence targets civilians. However, it is a particularly important question in modern stateless warfare, where it is a matter of which 'mode' of operations is being used by the insurgents. This issue is second only to 'type' in its analytical importance and arguably is second to none in its moral importance. So this analytical chapter focuses on the mode as well as type of insurgency before going on to provide a systematic and conceptual profile of insurgency as a distinctive type of non-state war.

The expeller, revolutionary and separatist types of insurgency

The types of insurgency are identified by the particular political goals the insurgents are pursuing, whether such transformational goals as the revolutionary overthrow of the state or such prosaic goals as 'persuading' the foreigners to 'go home'. Categorising them into three main, general types of insurgency has long been the conventional wisdom among analysts of insurgency, even if they differ in the way they label the three types.[1] Two of these types were implicitly labelled the 'separatist' and the 'revolutionary' in Chapter 1 but the third – insurgency against a foreign state – was referred to more in terms of the differences between nationalist, religious and localist examples than in terms of their similar opposition to being invaded, occupied or colonially ruled by a foreign state.

However, an appropriate label has been provided by an apt description of the Spanish guerrillas' landmark insurgency against Napoleon's army: 'the irregular fighter taking up any weapon at hand to *expel* invaders'.[2] The word 'expeller' seems an appropriate label for all insurgencies fighting to expel a foreign state that is invading, occupying or colonially ruling an insurgent's nation, religious community, local community or other traditionally or ideologically defined 'home'. The expeller type has been the most common of the three but only because of this type's prevalence among premodern insurgencies; it has been markedly less prevalent in modern stateless warfare despite its bewildering variety – including the Resistance movements against occupation by Nazi Germany, the anti-colonial insurgencies against French, British or Portuguese empires, and the jihadist Al Qaeda attempt to expel Western 'Crusaders' from the Middle East.

The separatist type of insurgency has some similarities with the anti-colonial expeller insurgencies, if only because the desire to separate from the local state and establish an independent state is similar to the anti-colonial demand for independence. Indeed such separatist insurgencies as the Basque ETA and the Provisional IRA have portrayed themselves as engaged in an anti-colonial struggle against a foreign rather than local state. However, it was pointed out in Chapter 1 that there is an objective basis or criterion (the nature of such political rights as representation in the national parliament) for making the distinction between separatist and anti-colonial insurgencies. Furthermore, the separatist type includes not only secessionist bands seeking a fully independent state but also those bands that are merely seeking some regional autonomy for their distinctive ethnic, sectarian or other social group. For example, after the Eritrean-Tigrayan separatist alliance's victory over the Ethiopian state, Eritrea became fully independent but Tigray opted instead for federal regional autonomy within a newly federalised Ethiopian state. And the separatist type is broad enough to accommodate the category of insurgencies that Clapham referred to as the 'warlord' type in his seminal study of African guerrilla movements.

These warlord insurgencies are characterised by the feudal-like goal of creating 'a personal territorial fiefdom separate from existing state structures and boundaries'.[3] So to help distinguish them more clearly from warlord semistates, they will be relabelled as 'neofeudal' rather than warlord insurgencies and be classified as the neofeudal subcategory of separatist insurgencies.[4] It is not surprising that such a broadly defined type can claim to be the most common manifestation of modern stateless warfare, despite the global prominence of the revolutionary type of insurgency in the 1940s–80s, and in fact the separatist type has clearly been more common than the revolutionary in the post-1945 era.[5]

Nonetheless, the public perception of insurgency still seems to be focused on the revolutionary type – on insurgents seeking to overthrow their local state and create a new form of state and society. Although the public perception of revolutionaries was focused upon the communists in the 1940s–80s, since then attention has shifted to the Islamist revolutionary insurgents and their desire to create some kind of Islamic state and society. In both the communist and Islamist cases there has been an international aspect to their revolutionary goals, as their ideology has envisaged 'the revolution' spreading to other countries and becoming a world-wide phenomenon. But it is doubtful whether much is to be gained from looking at the ultimate, global goals of Mao or Guevara or the Islamic ideologists. The more important international aspect of revolutionary insurgencies is the overlap with the expeller type that occurs when revolutionaries have to fight against both their local state and a foreign state that is assisting it, as when the Viet Cong were fighting against both the South Vietnamese state and the military forces of its US ally. In such cases the revolutionary insurgents are fighting a supplementary expeller insurgency whose success is a prerequisite or preliminary to achieving the revolutionary goal of destroying and replacing the local state. A contemporary example of this overlapping seems to be the Al Qaeda in Iraq (AQI) insurgents' fight against both their newly democratised state and the US military forces in Iraq. However, AQI and the 'headquarters' Al Qaeda are better described as being primarily expeller insurgents with supplementary revolutionary goals, as their anti-Crusader ideology and international recruiting are more indicative of an expeller than revolutionary insurgency. A better contemporary example would be the Taliban insurgency against the Afghan state and the Western military forces supporting it, but even in this case there is some ambiguity – not about expeller-versus-revolutionary priorities but about whether the Taliban's Islamist revolutionary goals have been subsumed by those of Pashtun ethnic/tribal separatism.[6]

All three types of insurgency have subcategories or versions, which have in turn produced a range of ideological variations. The communist version of the revolutionary type has been particularly fertile, with such varieties as Maoism and Castroism or Guevarism. The Islamist version, too, has a range of ideological variants, such as the Shiite Khomeinist and the Sunni Muslim-Brotherhood ideologies. The expeller type's two main versions are anti-colonial ideologies, and jihadist anti-communism and anti-westernism. Although the jihadist anti-communism became extinct with the expulsion of Soviet forces from Afghanistan, the jihadist anti-westernism has become the most important version of expeller insurgency – especially the anti-Crusader ideological variant popularised by Al Qaeda.

Table 2.1 Insurgency: Types, Versions and Bands

	War Against Local State	War Against A Foreign State
Types	a) Revolutionary and b) Separatist	c) Expeller
Typical Versions	a) communist revolutionary and b) ethnic separatist	c) jihadist expeller
Typical Bands	a) Viet Cong and b) Basque ETA	c) Al Qaeda

The guerrilla and terrorist modes of insurgency

Like categorising insurgency into three main types, it has long been the conventional wisdom to categorise insurgency into two different modes – the guerrilla and the terrorist. For example, more than 30 years ago Laqueur noted that the 'terrorist element has been more pronounced in some guerrilla movements than in others' and may be 'the predominant mode of the armed struggle'.[7] The two modes of insurgency, however, are related less to the three types than to the various strategies and operational formats used by insurgents, which are described in Chapters 3, 4 and 5. This is because the difference between the guerrilla and the terrorist modes is usually described in terms of the *targeting* of attacks and therefore is closely related to the strategy being pursued by the insurgents, such as a propaganda or provocation strategy, and may even be related to the operational format, whether rural bandit-like or urban gangster-like.

The distinction between the two modes can be summed up as whether the insurgent attacks are a) targeting the state's military forces or b) targeting the state's civilian citizens. Of course between these alternatives lies a large grey area, which includes a state's economic resources and its political leaders and state officials. However, that is a matter for further discussion below and for differing interpretations; the key point is that there is an obvious and important difference between the two modes. And of course an insurgency may well be operating in both modes, as when Osama bin Laden stated in his 1998 declaration of war: 'To kill the Americans and their allies – civilians and military – is an individual duty incumbent upon every Muslim in all countries'.[8] Yet even in these dual-mode cases there is usually a primary or fundamental mode that characterises an insurgency no less succinctly or significantly than its type does – whether expeller, separatist or revolutionary.

The term 'guerrilla' is the historically as well as linguistically appropriate label for the mode that is aimed at the state's military forces. Not only does the word mean a small or small-scale *war* but also it first appeared in history as a description of insurgents' attacks on a foreign state's invading army. As for the other mode, the obvious term to use is 'terrorist', even though it cannot be said to be a historically or linguistically appropriate label. In fact the term has been given so many different definitions that there is always a danger of conceptual ambiguity when applying it out of context.[9] Another, more important problem is that the term 'terrorist' has such strong pejorative connotations that it is difficult to apply the term in a politically neutral manner. Even before the 9/11 attacks Crenshaw was pointing to this problem and the difficulties faced by scholars who wanted to avoid the appearance of taking sides in a conflict that had produced accusations of terrorism.[10] What is more, the contrast between the two labels is heightened by the fact that the term 'guerrilla' seems to have quite favourable or positive connotations. For example, Kilcullen's recent theory of the 'accidental guerrilla' presented such insurgents in a more favourable light by describing them as guerrillas rather than insurgents – let alone terrorists.[11]

So to describe an insurgency as operating in terrorist rather than guerrilla mode may arouse a political controversy that will obscure important analytical and strategic issues. Nonetheless, the guerrilla-terrorist labels are the best available and also seem to sum up the popular perception of what distinguishes terrorism from other ways of going about stateless warfare or political violence. The guerrilla-terrorist labelling of the modes also seems to conform to the views of scholarly experts on insurgency. For example, Asprey's historical survey of guerrilla warfare described it as fighting 'against orthodox military forces'.[12] In his standard text on insurgency, O'Neill expanded Asprey's definition of guerrilla warfare to include attacks on police personnel and economic targets, but the guerrilla attack could still be distinguished from terrorism by the fact that guerrillas did not target unarmed civilians.[13] His category of 'unarmed civilians' did not distinguish between a) political and administrative leaders and b) ordinary civilians, who are not part of the government or regime. But many people would be reluctant to categorise the assassination of a dictatorship's political or administrative leaders as being acts of terrorism, particularly if the regime was genocidal or had conquered the assassins' country. Similarly, many people would want to distinguish between deliberate attacks on ordinary civilians and what O'Neill described as 'indiscriminate' attacks, in which the attacker is not concerned about any 'collateral damage' to ordinary civilians but does not view them as a target of the attack. Deliberate attacks on ordinary civilians are clearly more extreme than indiscriminate attacks and indeed seem to be at the extreme end of a targeting continuum or spectrum that extends through various categories of civilian targeting and ends at the opposite extreme with attacks on orthodox military forces (see Table 2.2). There is obviously an ambivalent or grey area in the middle that is open to interpretation – and perhaps also depends upon the context – but such lack of precision or uniformity is only to be expected when considering such a complex and contested issue.

However, the distinction between guerrilla and terrorist modes has long been complicated or confused by the apparently separate, cross-cutting distinction between a) rural insurgency and b) urban

Table 2.2 Targets of Guerrilla and Terrorist Modes of Operation

Military targets	Guerrilla
Police targets	Guerrilla
Economic targets	Guerrilla
Political leaders	Terrorist/Guerrilla
Administrative leaders	Terrorist/Guerrilla
Civilians: Indiscriminate	Terrorist
Civilians: Deliberate	Terrorist

insurgency. When the terms 'urban guerrilla' and 'urban guerrilla warfare' became prominent in the 1960s–70s, they were implying that the guerrilla mode of insurgency occurs in urban as well as rural environments. The theory and practice of urban guerrilla warfare became particularly prevalent in Latin America and to some extent superseded its rural equivalents.[14] But Laqueur's 1977 work on the theory and historical practice of guerrilla warfare argued that the term 'urban guerrilla' was normally being used as 'a euphemism for urban terrorism', which involves such activities as bank robbery, kidnapping and airplane hijacking rather than the guerrilla ambush 'in some remote jungle village'.[15] This argument is not only distinguishing between the urban gangster-like and the rural bandit-like operational formats but also contending that another difference between them is that the gangster-like format normally operates in the terrorist mode, that is, normally targeting civilians and their way of life.

A strategic explanation for this apparent difference was presented in Laqueur's companion book on terrorism, which pointed out that a (maoist-model) guerrilla strategy is not a viable proposition in urban areas. The urban environment 'dictates a different strategy' from the guerrilla leader's aim of 'building up ever-growing military units and eventually an army, and establishing liberated zones in which an alternative government can be put up'.[16] He acknowledged that guerrillas pursuing this (maoist-model preparation) strategy in rural areas have 'used terrorist techniques on many occasions' but he contended that the reverse did not apply – urban insurgents had not used guerrilla techniques on many occasions – and reiterated that an urban environment was unsuitable for guerrilla warfare.[17]

Such strategic explanations, however, should take into account all the alternatives to the maoist-model preparation strategy, especially the pressure-strategy alternative. Unlike the typically terrorist propaganda and provocation strategies, the pressure strategy is not typically associated with either mode – guerrilla or terrorist. It is true that history suggests a revolutionary insurgency will not succeed in defeating its local state by relying on the pressure strategy. However, when attention is shifted from the revolutionary to the expeller type – and from Latin America to the Middle East – pursuing a pressure strategy in an urban environment appears to have some chance of success, even when its gangster-like operational format is not operating in the terrorist, civilian-targeting mode. There is the historically famous example of Begin's Jewish-nationalist Irgun band and its 1940s expeller insurgency against British rule in Palestine. Although it used a gangster-like operational format, its attacks targeted the British military, police, administrative and economic presence, not ordinary civilians, and this was epitomised by the time-bomb attack that blew up the British administrative secretariat and army headquarters housed in Jerusalem's King David Hotel.[18] Some 60 years later, there is the contemporary example of the Iraqi Sunni insurgent bands that have used a pressure strategy, and an urban gangster-like operational format, in an attempt to expel the US from Iraq. Although they have not been as strategically successful as Begin's Irgun, they have inflicted far more casualties on US forces than his band was able to inflict on the British forces stationed in Palestine.

On the other hand, the British and US forces did not refer to these urban insurgent foes as 'guerrillas'. And the reluctance to describe urban insurgents as guerrillas is also evident in the case of the third, separatist type of insurgency – in a European rather than Latin American or Middle Eastern context. When O'Neill's text on insurgency was discussing the possibility of an urban form of guerrilla warfare, it noted that the Basque ETA and the Provisional IRA were examples of urban insurgencies that in terms of their targeting were more akin to guerrillas than terrorists. He pointed out that when urban groups carry out 'bombings of military barracks and vehicles and sniper attacks on police and military officers', these acts are 'more akin' to guerrilla warfare than to terrorism and would be readily categorised as guerrilla operations if carried out in a rural environment.[19] And he noted later that 1) the (Provisional) IRA were an example of a group that carried out urban guerrilla-like attacks as well as indiscriminate bombings that have killed unarmed civilians and 2) the Basque ETA were an urban group that primarily targeted military and police personnel – which implies that they should be described as primarily guerrillas or as primarily using a guerrilla mode of insurgency.[20] Yet these insurgents were referred to as terrorists by their foes, and even their supporters have not described them as urban guerrillas. So it is hardly surprising that O'Neill's analysis of insurgents' urban-warfare strategy did not mention urban guerrillas and referred only to urban terrorism and urban insurgents.[21] The notion of an urban guerrilla mode of insurgency became anachronistic after the 1970s and the most that urban insurgents can hope for by using a guerrilla rather than terrorist mode in their targeting is to be labelled 'insurgents' or 'fighters' or 'militants' or 'rebels' but not 'terrorists'.

The guerrilla/terrorist distinction has been further complicated by the two different legal-institutional perspectives on insurgency that arise from the division of labour within the state between civilian and military means of countering insurgents. While civilian police, intelligence, and judicial institutions counter 'terrorists' according to criminal law, military institutions counter 'insurgents' according to military law and international conventions relating to counterinsurgency conflicts. A Clausewitzian strategist has pointed out that 'just because terrorists have to be countered principally by police forces and intelligence agencies, and are brought to trial for their crimes, we should not be confused about the nature of their activity. Terrorism is warfare and terrorists are soldiers'.[22] But the key point is not the incongruity of categorising an insurgency as criminal rather than military simply because it is small enough to be countered by civilian rather than military means; the key point is instead the incongruity of categorising an insurgency as 'terrorist' without considering whether it is operating in a terrorist rather than non-terrorist mode of stateless warfare.

What is more, from these legal-institutional perspectives a state seems to be acknowledging that it is fighting against insurgents, not terrorists, if it takes a military rather than civilian approach to dealing with the problem. In particular, the US declaration of a 'war on terror' in response to the 9/11 attacks appears to have set a global standard or norm of viewing such conflicts as a form of warfare rather than criminality. It is true that as early as 1984 Secretary of State Schultz had described terrorism as a new form of warfare, and President Reagan had directed that military options be prepared to deal with it, but this declaration of war against an unspecified terrorist foe had been referring to state-sponsored terrorism.[23] And the first military response or retaliation, the 1986 air attacks on Libya, was directed against a state believed to be sponsoring terrorist attacks on US citizens. In contrast, the US government's response to the 9/11 attacks was directed primarily against the Al Qaeda insurgent band and only secondarily against the Taliban semistate that was sheltering them but the response took the form – in rhetoric, law, and countermeasures – of a military response. A British historian of war quickly pointed out that the declaration of a war against terrorism was 'a very natural but terrible and irrevocable error' that could be contrasted with British governments' use of the term 'emergency' to describe situations in which the police and intelligence services were given exceptional powers and 'reinforced where necessary' with military personnel yet still treated the terrorists as criminals rather than belligerents.[24] However, these British governments had been fighting against separatists in Ireland or expeller insurgencies in territories under their colonial rule; they had not been fighting against a third-phase expeller insurgency sheltering in lands not under their rule and perhaps not under any state's rule.

Distinguishing insurgency from other non-state warfare

Systematically distinguishing the various types and modes of insurgency highlights the need to distinguish insurgency itself from the various other forms of non-state warfare – and to make this distinction in a systematic and conceptual fashion. For example, insurgency can be distinguished from civil war by specifying that the latter is a civil conflict involving one or more *semistates*, not one or more stateless bands. Of course semistates fighting against foreign states will have to be categorised separately as fighting 'expeller' wars against an invading, occupying or colonial foreign state. Furthermore, some civil conflicts between semistates or between a semistate and its local state seem too small to be regarded as a civil 'war', but they can be excluded by using the standard quantitative criterion of requiring a certain minimum level of casualties, such as the 1000-deaths-per-annum threshold.[25]

This structural, stateless-semistate distinction between insurgency and civil war in turn begs the question of what exactly is a semistate and how does it differ from an insurgent or other stateless band. Unlike a stateless band, a semistate is a territorial political entity that is a flawed version of the modern territorial state. The semistate rules over a sizeable part of a state's territory but this rule, even if based upon a monopoly of the legitimate use of force, is not internationally recognised as legally legitimate. The semistate is thus in a sense the opposite of what Jackson refers to as a quasi-state, which is a state that is internationally recognised as legally legitimate but does not exercise effective rule over its territory.[26] A semistate is very similar to a 'de facto state' but the qualifier 'semi' is a reminder that the effective rule is being exercised over only part of the territory of an internationally recognised state.[27] More importantly, the 'semi' highlights the fact that even the most effective semistate is still a severely flawed example of a modern state. As Mao said of the communists' rural base areas, 'the form of our political power is still far from that of a full-fledged state'.[28]

A relatively recent and well-known example of a semistate engaged in civil war is the Bosnian Serb semistate's activities in the Bosnian civil war of 1992–5, which produced the term 'ethnic cleansing'. Even before the Bosnian Serbs began the opening offensive of their separatist war they had been setting up 'Serbian autonomous regions' (SAO) all over the parts of Bosnia in which the Serb minority had a territorially definable presence: 'proclaiming an SAO, or *microscopic statelet* ... seemed ridiculous. There were jokes about the SAOs, but checkpoints manned by Serb gunmen drove home just how serious the Serbs were'.[29] Soon the Bosnian

Serbs began extending their territorial holdings and linking up SAOs into larger areas. What is more, they also began evicting and sometimes even eliminating any non-Serb inhabitants of these territories – a process that was described as 'the cleansing of the ground' but would later became known as ethnic cleansing.[30] Thus the newly declared Bosnian Serb state was given 'territorial definition' by the 'Serb paramilitary hit-squads' that had taken control of these areas.[31] Although these Serb paramilitary squads were structurally more like insurgent bands than regular armies, they were fighting a semistate war of taking and holding – and 'cleansing' – territory that would be used to expand the area under their semistate's rule.

In contrast, warfare between stateless bands lacks this territorial and political/ruling element, displaying its primeval origins as the prestate form of warfare, but it can have as high a 'body-count' as a civil war. For example, the rightist versus leftist inter-band conflict in Turkey in the later 1970s had a death toll that eventually reached 'the oft-quoted figure of 20 victims a day' and no fewer than 1,500 were killed in the months preceding the 1980 takeover by the military – which was welcomed by a nation 'on the verge of collapse and civil war'.[32] A less bloody but more complicated inter-band conflict had arisen in Northern Ireland in the early 1970s. An extra complication is that the inter-band violence between the IRA and such Protestant bands as the UVF was occurring within the context of the IRA separatist insurgency against the British state. To add to the complexity, there was also inter-band conflict between the Provisional IRA insurgents and the Official IRA insurgents, who were rivals for the support of the Catholic sectarian minority while still being separately engaged in conflict with Protestant bands and the British state's forces. The complexity of the situation was illustrated by a brilliant 1974 cartoon of a British soldier in Northern Ireland ruminating about the bullets flying around him:

> Well it could be the Officials [IRA] firing at the Provos [IRA], or the IRA firing at the Prots [Protestants], or the UDA [Ulster Defence Association] firing at the Officials, or the UVF [Ulster Volunteer Force] firing at the Cats [Catholics] – on the other hand, they could all be firing at me [the British army].[33]

A pyramid seems the best geometric shape to represent the multisided combination of a) vertical insurgency with b) horizontal inter-band conflict.

However, there is no need to categorise the pyramidal, multisided insurgency as a separate type or kind of war from the standard insurgencies. It is quite a different case from having to distinguish between a civil war and an expeller war because semistates' wars against foreign rather than local states are clearly not 'civil' or 'internal' and therefore must be given a different label (see Table 2.3).

Table 2.3 Types of War (stateless warfare italicised)

	Against one or more bands	Against a semistate	Against a state
By a state	i) counter-insurgency	ii) civil or expeller war	iii) 'normal', inter-state war
By a semistate	i) counter-insurgency	ii) civil war	iii) civil or expeller war
By a band	i) *inter-band conflict*	ii) *insurgency*	iii) *insurgency*

The vertical nature of insurgency is what best distinguishes it from the other non-state forms of warfare. The verticality that is present in the term's original Latin meaning of 'rising up' against a state is also very evident when the structure of insurgent stateless warfare is compared to the structure of other non-state wars. For wars 1) between stateless bands and 2) between semistates are structurally similar to 3) wars between states, because in all three situations they are wars between what Waltz famously described as 'like units'. In his theory of international relations he depicted states as being 'like units' in the sense of having similar tasks or functions, even though they differ in their ability to perform these activities – they have 'varied capabilities' that range from a superpower's to those of a weak Third World state.[34] If the relationship between two states descended into warfare, it would still be a relationship between two 'like units' with different capabilities but with an emphasis now on their capability to wage war. The same analysis can be applied to wars between *non-state* 'like units'. Instead of wars between states there are 1) inter-band conflicts between stateless bands with different military capabilities or 2) civil wars between semistates with different military capabilities. In contrast, an insurgency waged by a band against a state is a war between *un*like units that differ in their form of capability, not simply their level of capability, to wage war against each other. Structurally, this contrast can be depicted as another expression of the verticality of insurgency – of rising up against an unlike unit – rather than the horizontal nature of wars between 'like units' that 'square off' against each other.

It is true that an insurgency against a state is not the only kind of war that has some verticality – some 'rising up' against an unlike unit. Clearly this structural feature also occurs in cases of civil or expeller war between a semistate and a local or foreign state (see Table 2.3). But they do not have such a starkly vertical structure as insurgency against a state, because the adversaries are not so unlike each other. As was described earlier, a semistate is similar to a state territorially, politically and militarily and in fact expeller or civil wars between these two 'unlike' units have often seemed quite horizontal in nature – comparable at least militarily to wars between semistates or states.

On the other hand, the historically few and also unimportant examples of a band fighting against a semistate are clearly vertical in structure and are categorised as a kind of insurgency (see Table 2.3). Although a stateless band has more in common with a semistate than with a state, the band is 'rising up' against a unit that is unlike it territorially, politically and especially *militarily*. Any insurgent band has a very different 'way of war' from the semistate's and state's regular or conventional warfare, and in fact an insurgency's way of war is as starkly vertical as the structure of a band's insurgency against a state.

Insurgency's 'way of war' seems so vertical in nature because the band appears to be operating in a different spatial dimension or medium than the state's forces – and therefore to be making its attacks in a different, vertical direction or plane. In Clausewitz's chapter on what he called 'people's war', he noted that 'the element of resistance will exist everywhere and nowhere'; it should be 'nebulous and elusive' but at certain points 'the fog must thicken and form a dark and menacing cloud out of which a bolt of lightening may strike at any time'.[35] His use of the French aphorism that the enemy is everywhere and nowhere would be repeated by Giap in the 1950s when he pointed out that the French forces fighting against his Vietnamese insurgents were faced with the problem of 'where is the front?' and 'the enemy is everywhere and nowhere'.[36] Giap's emphasis on the *front*less nature of this stateless warfare highlighted the fact that insurgency was operating in a different dimension or medium than states and semistates attacking each other horizontally from their 'front lines'. Warfare between bands, too, is horizontal because the bands operate in the same dimension or medium as their enemy and therefore attack each other in a horizontal direction or plane – even if, unlike states and semistates, they attack each other in a frontless way.

The vertical nature of insurgency's way of war clearly distinguishes it from inter-band and semistate forms of non-state warfare – and in fact it has more in common with the vertical aspects of modern inter-state warfare. In modern times some states have developed two technological forms of vertical warfare, attacking from aerospace and from beneath the sea, which are beyond the means of a band or semistate to develop or acquire but have some important similarities with insurgency's way of war. These two technological forms of vertical warfare became increasingly important during the 20th century, and it might well be argued that by the end of the century the vertical way of war dominated inter-state as well as stateless warfare. Although the First World War's trench warfare on the Western Front and battleship salvoes at Jutland had epitomised the horizontal attack, the airplane and submarine had already been showing the potential of attack from above and below. That potential was realised in the Second World War and was boosted to new heights by new technological developments in the 1940s–50s. Attack from the air or aerospace became apocalyptically devastating with the advent of nuclear bombs and missiles, while the vertical attack from below the sea developed into a combination of the two forms when nuclear-powered submarines were equipped with nuclear missiles. The first submarine armed with Polaris intermediate-range nuclear ballistic missiles was deployed as early as 1960, and seven years later the US had more than 40 of these submarines – together able to attack as many as 300 cities in the Soviet Union.[37]

Like an insurgent band, these airplanes, missiles and submarines attacked in a vertical rather than horizontal direction and were operating in a different, third dimension. A theorist of air warfare has recently described how the 'effective utilization of airpower added, for the first time in history, a *third dimension* to war: a dimension which allowed direct attack on enemy rear zones, cities, economies and, perhaps most importantly, civilian populations' – the 'zone of conflict was extended' to far behind the front line.[38] However, a third dimension of war had in fact existed ever since the effective utilisation of insurgency in premodern times, even if these insurgencies were seldom able to extend the zone of conflict very far behind the front line.

Insurgency is better compared to attack by submarine, though, and indeed this comparison was made as long ago as the 1960s when Huntington compared guerrilla warfare to the classic naval strategy of the *guerre de course* (the war of raiding).[39] This strategy of raiding the enemy's seaborne trade and supply – the 'commerce-raiding' sinking or capturing of merchant ships – aimed to weaken the enemy's economy and political commitment to the war:

> Its weapons were the frigate and *privateer* in the eighteenth century, the torpedo boat and cruiser in the nineteenth century, and the *submarine* in the twentieth century. Yet the tactics of the *guerre de course*

remained essentially the same. They are also the tactics of guerrilla warfare. ... Both employ the tactics of hit-and-run. The security that the guerrilla finds in the rugged terrain and a friendly populace the commerce raider finds in the empty reaches and lower depths of the ocean.[40]

Some of these historical comparisons are not exactly new. As was described in Chapter 1, the Spanish guerrillas of 1808–9 were initially described as land privateers, while the first commerce-raiding campaign by submarines, instituted by Germany in the First World War, led to the submariners being compared to pirates by their foes and being described as 'undersea raiders' in a celebrated post-war account of their exploits, *Raiders of the Deep*.[41] But Huntington seems to have been the first to make the direct comparison between submarines and guerrilla bands, despite it having been foreshadowed decades earlier by Mao's famous comparison between fish in water and guerrillas operating among a supportive population.

The appropriateness and usefulness of the comparison is confirmed by historical analyses of the strategy and tactics of submarine warfare. For example, Lautenschlager's magisterial analysis points to the submarine's distinctiveness in operating in a separate, underwater 'medium' from surface ships and therefore having distinctive tactical strengths and weaknesses when carrying out torpedo attacks on surface ships.[42] On the one hand, submarines are difficult to find except when using their weapons or active sensors and they operate most capably in hit-and-run surprise attacks that give them the advantage of attacking first; on the other hand, they depend upon stealth for their survivability and their capability for surprise attacks, they depend upon hitting with their first shot/salvo of torpedoes and they carry quite a limited supply of these weapons (which seem similar to such insurgent weapons as a suicide-driver car bomb). Considering how much the submarine's survivability and capability depends upon hiding underwater, it has been aptly described as carrying out not only hit-and-run but also 'hit and hide' attacks and in fact the most appropriate label might well be *hide-and-hit* attack.[43]

The technological developments of the 1940s–50s led to a submarine with a nuclear-powered ability to hide underwater and a nuclear-armed ability to hit – to deliver an apocalyptic attack on the enemy state. For 'when ballistic missiles armed with nuclear warheads were deployed aboard submarines', they were given the capability 'to destroy the population centers, industrial capacity, or economic infrastructure of a state'.[44]

Insurgents have not yet experienced such a revolution in their ability to hit and power to hurt but, as the Conclusion suggests, new technology may eventually provide them with an equivalent of the submarine's nuclear ballistic missiles. Until that technological revolution in insurgent capability, insurgency will be tactically and strategically comparable to only the pre-nuclear submarines, with their hide-and-hit torpedo tactics and 'war of raiding' strategy. As Huntington pointed out, the pre-nuclear submarine was not very successful; in both world wars it had been no more than a 'supplementary weapon' or a 'harassing device'.[45] In contrast, insurgency has won some major victories with its version of the war of raiding, which is described in the next chapter as a pressure strategy and compared to the other main insurgent strategy – the maoist-model preparation strategy.

3
Insurgent Strategies

Unlike the previous two chapters' historical and conceptual analyses of insurgency, this chapter is not identifying and describing 'the nature of the enterprise' – it is focused on how the insurgents intend to make it a *successful* enterprise. It is concerned with how they plan to use organised violence to achieve their political goals or, in other words, the 'strategies' they use to win the war or a major campaign in the war rather than merely a particular battle or a series of small-scale military actions. Insurgents' various strategies can be categorised under four headings, namely preparation, pressure, propaganda and provocation, and can also be related to other aspects of insurgent stateless warfare, especially to its historical phases, its three main types and its modes of operation.

Thus an underlying theme of this chapter's description of the two main strategies, preparation and pressure, is that the maoist-model preparation strategy produced by the second phase of modernisation seems to have become outmoded, particularly when compared to an alternative strategy of exerting indirect political pressure on the enemy state. However, there is also an appraisal of how the three types of insurgency, revolutionary, separatist and expeller, have used the preparation strategy – or the pressure strategy instead – and there is an extensive discussion of how the pressure strategy indirectly exerts its political pressure through the military power to hurt and the 'political calculus' that this produces. These two key strategies are included in the next chapter, too, as it examines how the terrorist mode of operation has been used in conjunction with all four insurgent strategies but especially the typically terrorist strategies of propaganda and provocation.

Preparing for battle – the maoist model

The notion of strategies and strategic thinking as planning how to 'win the war' is difficult to apply to stateless warfare until the 20th century. Earlier strategic thinking and writings, notably Clausewitz's, were concerned with insurgency from the perspective of warfare between states, such as how a state could use nationalist guerrillas to defeat an invading army. In contrast, Mao Zedong's writings on guerrilla warfare presented strategies from the perspective of the insurgents, as in his 1936–8 publications: 'Problems of Strategy in China's Revolutionary War', 'Problems of Strategy in Guerrilla War Against Japan' and of course his famous 'On Protracted War'. Even as early as 1929 his 'On Correcting Mistaken Ideas in the Party' contained a criticism of the notion of relying on 'roving' guerrilla bands roaming the countryside rather than building up permanent base areas.

As was described in Chapter 1, Mao's theory of guerrilla warfare focused upon preparing for a shift to semistate warfare which would destroy the local state through battles and campaigns that took and held its territory. His preparation strategy was therefore focused on establishing large rural base areas – semistate areas – in which the guerrillas could create a regular army that was capable of defeating the state's forces on the battlefield. Thus the maoist-model strategy of guerrilla warfare can be summed up as preparing to fight large-scale battles against the state's army.

In other words, the maoist-model insurgents are preparing to fight the state's army on its own terms – an inherently dangerous and risky approach. When the maoist-model regular army has been fully prepared, apparently it will take on the state's army with the same tactics and strategies of 'normal' warfare that the state's army has been professionally trained and technically equipped to defeat. But a recent 'theory of asymmetric conflict' has pointed out that the weaker side in any form of conflict is likely to win only if it is using a *different* strategic approach from that of the stronger side; if both sides are using the same strategic approach, obviously the weaker side is likely to be defeated.[1] And the foolhardiness of an insurgency giving the enemy army something army-like to fight against was noted by Clausewitz when he warned that an uprising against an invading army 'should never materialize as a concrete body, otherwise the enemy can direct sufficient force at its core, crush it, and take many prisoners'.[2] In fact it is very difficult merely to go about preparing for a semistate war without attracting a devastating preemptive attack on the base area in which the insurgents' regular army is being recruited, equipped and trained.

So it is hardly surprising that revolutionaries have seldom had much success with the maoist-model's preparation strategy, and on the few occasions they have been victorious it has been due to a factor – foreign or urban assistance – that is not part of the maoist-model strategy. After all, Mao's theory of guerrilla warfare involves creating regular armies in rural base areas and then 'marching on the capital' and the other major cities; there is no mention of assistance from revolutionaries in the urban areas – let alone assistance from foreign countries. Yet the historical record shows that maoist-model revolutions have always required assistance from a foreign state (deliberate or unintended) or from the urban population of the insurgents' society. In the language of 'normal' warfare, it has required 1) foreign allies or diversions and/or 2) a second, urban front that complements the band's rural activities.

Even Mao's successful use of the preparation strategy was dependent on foreign allies and diversions. It was only the onset of the 1937–45 Sino-Japanese War that enabled the communists to establish the extensive base areas and large regular armies that in turn later enabled them to win the civil war against Chiang Kai-shek's Nationalist regime. The Japanese armies occupied vast areas of China in the early years of the war but lacked the manpower to control such huge expanses of territory, which gave the communists the opportunity to establish extensive base areas behind Japanese lines and create a communist semistate that incorporated 19 base areas, more than 40 million people and hundreds of thousands of soldiers.[3] In addition to these crucial benefits, the war against Japan led to the communists' foreign ally – the Soviet Union – being able to supply them with crucial assistance. After the Soviet Union joined the war against Japan in August 1945, Soviet armies quickly overran the Japanese-ruled northeast of China and began to supply the Chinese communists with not only military expertise but also captured Japanese heavy weapons and large stretches of formerly Japanese-held territory in the northeast.[4] With this assistance the communists' regular army, the People's Liberation Army, continued its phenomenal growth path that had already seen it expand from only 90,000 soldiers in 1937 to some 475,000 in mid-1945 and would now see it reach nearly two million by 1947, when the civil war with the Nationalist regime was still largely confined to the northern regions of China.[5]

Some later examples of successful maoist-model revolutionary insurgencies also depended upon foreign allies and their assistance. This foreign assistance often went beyond supplying equipment and training to include providing cross-border sanctuaries (base areas located across the border in the friendly state and therefore safe from the enemy state's attacks) and even sending troops to fight alongside the insurgents as fellow guerrillas or as regular forces participating in a shift to semistate warfare. The most well-known examples are the communist takeovers of South Vietnam and Cambodia in 1975. The Viet Cong's dependence upon their ally North Vietnam was symbolised by the cross-border invasion that finished off South Vietnam, while the Cambodian Khmer Rouge insurgents were wholly dependent upon North Vietnamese troops during the crucial years of 1970–1, when the Khmer Rouge were acquiring huge liberated zones and creating the regular army that would deliver final victory in 1975.[6]

The other cases of successful maoist-model revolutions have instead benefited from a second, urban front and have also been fighting against a small country's weak state. The Castro-led Cuban revolution is the most famous of these cases but post-revolutionary Cuba created a 'myth of the heroic guerrilla' that has obscured the important role played by the second, urban front in helping Fidel Castro's Rebel Army establish its Sierra Maestra base area in 1956–7 and march upon the cities in 1958.[7] Research in Cuban archives has confirmed that the urban wing of Castro's 26 July Movement (M267) played a crucial role in supplying recruits and logistical support to their rural counterparts fighting a guerrilla war in the mountains in 1956–7.[8] And it has shown that Castro's two-front, rural-urban strategy initially placed greater emphasis on the urban front, which was intended to combine workers' strike action with armed attacks by the M267 underground militia.[9] The April 1958 disastrous failure of the urban militia attacks and general strike forced the M267, almost literally by elimination, to re-orient the revolutionary strategy towards the rural Rebel Army. If Debray is correct in suggesting that the guerrillas independently invented a maoist-model strategy, it was not until this late stage in the guerrilla campaign that necessity became the mother of invention. Apparently the 'rebels discovered the writings of Mao' only at the end of the insurgency and 'found in this book what they had been practicing under the pressure of necessity', but he acknowledges that not until April 1958 did the Rebel Army 'set up a firm guerrilla base in the centre of the Sierra Maestra'.[10] When the Rebel Army began its march on the cities later in that year, it still numbered only a few thousand troops and was more a guerrilla band than a regular army; it was very fortunate that its foe was a small and weak state with very poor quality military forces.

A similar two-front, rural-urban combination was used by the Sandinista revolutionary insurgents in Nicaragua in 1979. They were facing a much smaller army of only about 10,000 soldiers, as compared to over 40,000 in the Cuban army faced by Castro, but the Nicaraguan army was far more professional and effective than its Cuban counterpart had been in the 1950s. Thus the Sandinistas' small rebel army in turn had to be more regular in organisation and equipment than its Cuban counterpart had been and its final offensive had to include an urban insurrection – involving a general strike, urban militia insurgents and heavy fighting in the cities – as a means of tying down the state's military forces and taking pressure off the rurally based rebel army.[11]

The importance of foreign allies and/or urban fronts is also displayed by two cases in which the revolutionaries failed to seize power and had to be content with a second-best peace agreement. Unlike their counterparts in Nicaragua, the FMLN insurgents in El Salvador were unable to establish an urban front. In 1981 an offensive by their rebel army almost reached the capital city but did not provoke an urban insurrection, and in 1989 a large-scale offensive by the now weaker insurgent forces did win some support in working-class districts of the capital but could not provoke a mass insurrection.[12] Realising that it had no answer to the military stalemate, the FMLN signed the 1992 Peace Accords that transformed it into a peaceful political party in exchange for UN-supervised military reforms and security guarantees by the state.[13]

A similar and more recent case is that of the maoist-model insurgency by the explicitly Maoist communist movement in Nepal, the Communist Party of Nepal (Maoist). The insurgents' success in 'liberating' the countryside and building a People's Liberation Army did not lead to a march on the capital city because not only was the Nepalese state still able to prevent an urban insurrection in the capital but also the insurgents believed that any such urban insurrection would lead to intervention by foreign troops and an 'Afghanistan-like situation'.[14] Consequently, the insurgents were prudently content with a compromise peace agreement, the 2006 Comprehensive Peace Agreement, that required both their PLA and the state's army to be confined to barracks and incorporated the Maoists in a political process of democratisation through a Constituent Assembly.[15]

Somewhat paradoxically, the only example of a maoist-model strategy achieving complete success *without* foreign or urban support is a case of separatist, not revolutionary, insurgency – which succeeded in winning independence for the Eritrean region of Ethiopia. This long separatist struggle of the 1960s–90s did not establish an urban second front or receive much assistance, deliberate or unintended, from foreign states. And it was fighting against one of Africa's most powerful states, which eventually mobilised military forces numbering in the hundreds of thousands and equipped with heavy weapons and air support provided by the Soviet Union.

The Eritrean separatists had certainly fought what Mao termed a 'protracted war', which in this case took some 30 years to achieve victory.[16] By the later 1960s they were staging an effective campaign of rural guerrilla warfare, and in the mid-1970s they shifted successfully to semistate warfare against an Ethiopian state that was undergoing the turmoil of a military-led Marxist revolution and an unsuccessful war with Somalia. The separatists took and held most of Eritrea until a 1978 counter-offensive by a much strengthened, Soviet-equipped Ethiopian army forced them to give up their territorial gains and retreat northwards to the mountainous Nakfa base area. There they repelled a series of offensives by the Ethiopian military and in 1982 won a remarkable victory when the 22,000-strong army of the Eritrean People's Liberation Front counterattacked more than 80,000 Ethiopian troops equipped with Soviet-supplied heavy weapons and air support.[17] Reinforced with tanks and other captured equipment, the EPLF eventually began an offensive in 1988 that led to the victorious battle of Af Abet and the opening of the way south to the major cities of Eritrea.[18] By 1991 it had taken most of these cities and was helping its Tigrayan-separatist allies complete the final drive towards the Ethiopian capital of Addis Ababa.[19]

A better-known but less successful separatist maoist-model insurgency was fought against the Sri Lankan state by the Tamil-separatist Liberation Tigers of Tamil Eelam (LTTE) or 'Tamil Tigers'. Their protracted war was apparently ended by the Sri Lankan state's 2008–9 military campaign, which eliminated the Tamil Tigers' semistate and their leaders. By then the Tigers' separatist struggle was in its third decade and since 1983 had experienced a series of four different phases or separate wars interspersed by ceasefires, notably the 2002 Ceasefire Agreement that had enabled the Tamil Tigers to consolidate a large semistate known as 'Tamil Eelam'.[20]

In fact the Tamil Tigers seemed to have developed a new, separatist model of preparation strategy. This model sought to establish a semistate base area as virtually an end in itself, not as merely preparation for shifting to semistate warfare and destroying the Sri Lankan state by taking and holding its territory – including the capital city – like the Eritrean separatists' destruction of the Ethiopian state. The Tigers' strategy was to establish a semistate incorporating all or much of the territory inhabited by its ethnic minority and then, if necessary, fight a defensive semistate war that indirectly would politically pressure the Sri Lankan state into accepting the secession of this territory. In a sense the strategy was similar to that fought by the Confederate separatist semistate in the American Civil War:

> The Confederacy waged a strategically defensive war to protect its territory from conquest and preserve its armies from annihilation. To 'win' that kind of war, Confederate armies did not have to invade and conquer the North: they needed only to hold out long enough to force the North to the conclusion that the *price* of conquering the South and annihilating its armies was *too high*, as Britain had concluded in 1781 [and two years later had signed a treaty ending the American War of Independence].[21]

Such cost-benefit calculations tend to be made by the stronger side in a civil war because it is fighting more of a limited war rather than a war of survival. And if the costs of continuing the war seem too high, political pressure will be exerted on the government to end the war by negotiating a peaceful settlement of the conflict. In the case of the American Civil War, three of General Robert E. Lee's military campaigns 'each came close to sapping the Northern will to continue the war' and as late as August 1864 'Lincoln came under enormous pressure to negotiate peace with the Confederacy'; his refusal to bow to this pressure seemed likely to cost him any chance of re-election in November and it was only unexpected battlefield victories that 'turned Northern opinion from deepest despair in the summer to confident determination by November'.[22] If Lincoln had either succumbed to the pressure or been replaced by a more pliable President, this would have been a classic example of a semistate rather than insurgent band successfully pursuing the pressure strategy against a state.

In comparison, the Tamil Tigers were pursuing a two-stage version that was a variant of the insurgent preparation strategy but one in which the insurgents were preparing for a shift to a semistate *pressure* strategy. By the time of the 2002 ceasefire the Tigers had won control of large areas of territory in the Tamil region of the country. Their success was partly due to foreign assistance from one of India's federated states, Tamil Nadu, which was home to some 60 million ethnic Tamils and provided a 'logistical base and sanctuary' that was separated by less than 30 kilometres of sea from the Tamil region of Sri Lanka.[23] However, the Tigers had also shown a high degree of commitment to their cause, exemplified by the suicide attacks discussed in Chapter 4, and on several occasions domestic and/or international political pressure on the Sri Lankan government had led to a ceasefire and peace negotiations. The 2002 Ceasefire Agreement went further and tacitly recognised the de facto autonomy achieved by the insurgents' Tamil Eelam semistate. During the several years of peace that followed the Ceasefire, the LTTE was 'in full control of large areas, especially in northern Sri Lanka':

> Travelling from government-controlled to LTTE-controlled areas resembles a border crossing between two nation-states, with well-guarded border control posts where travellers are required to show identity cards, goods are inspected and customs fees are collected. Within the areas they control, LTTE runs a *de facto* administration, which includes revenue collection, police and judiciary as well as public services and economic development initiatives.[24]

Even during the 1990s the LTTE had established state-like control of the territories under its control. It had promulgated criminal and civil legal codes, established a judiciary and a police force and sought to collect direct and indirect taxes.[25]

However, the Ceasefire had not formally recognised the secession of the Tamil Eelam semistate and its right to be considered an independent state – and no country or international organisation recognised it as an independent state. Nor had the Tigers ever been willing to accept a second-best separatist option of federal or quasi-federal autonomy within the Sri Lankan state. Thus the Ceasefire was not a permanent negotiated settlement of the conflict and, like the other ceasefire periods in this long-lasting insurgency, its only temporary nature would confirm Zartman's pessimistic analysis of civil wars – and insurgencies of a similar nature. In the early 1990s he described them as often producing neither a conclusive victory nor 'a costly deadlock driving the parties to negotiate', which he had termed a 'mutually hurting' stalemate that through 'cost/benefit calculations' leads the two sides to reach a negotiated resolution of the conflict.[26] Instead there was often 'a stable and supportable' stalemate that generally involves an *un*recognised partition of the state's territory between the two sides.[27] In the Sri Lankan case, the Ceasefire Agreement had produced a tacitly recognised but nonetheless formally *un*recognised partition of the state's territory that was not much better than a stable and supportable stalemate. It was the equivalent neither of the negotiated resolution of the conflict in El Salvador through the 1992 Peace Accords nor of the conclusive separatist victory that created the independent state of Eritrea.

But in 2009 a conclusive anti-separatist victory was achieved by the Sri Lankan state, with a large-scale military offensive destroying the Tigers' semistate and eliminating its leaders. It made a striking contrast with not only the Eritrean situation but also the situation in El Salvador, where the formerly insurgent FMLN won the 2009 presidential election in a way which seemed to epitomise the successful negotiated resolution of the conflict. And the Sri Lankan state's conclusive victory also provided a striking illustration of how dangerous it is for insurgents to adopt the Tamil Tigers' variant of the preparation strategy. For even a defensive, pressure strategy means competing militarily against a state on its own terms, and the rapid military collapse of the Tamil Eelam semistate confirmed that this is much more difficult in the 21st century than it was in the 19th century era of General Robert E. Lee.

Usually it is even more difficult for insurgents to compete against a *foreign* state on its own terms, as expeller insurgents have found to their cost whenever they have shifted to semistate warfare. However, Mao's most famous description of the shift from guerrilla to regular warfare, the 1938 article 'On Protracted War', was presented in terms of an expeller, not revolutionary, insurgency – expelling the Japanese invaders and occupiers of China. He envisaged that the 'three-act drama of China's War of Resistance' would pass through the stages of 1) guerrilla warfare and the establishing of base areas, 2) continuing guerrilla warfare but supplemented by regular armies' mobile warfare and 3) taking territory through regular armies' mobile warfare and their supplementary positional form of warfare.[28] In practice this three-stage process never occurred in China and it would be the country's US and Soviet allies that defeated Japan and ended its occupation of much of China.

The nearest thing to a three-stage 'war of resistance' was fought in Yugoslavia, where Tito's communist Partisans eventually expelled Nazi Germany from the country that it had invaded and occupied in 1941. The small communist party was able to exploit the country's mountainous terrain to grow into a substantial guerrilla force that in 1942 went a stage further by creating a regular army (the JANL) and setting up local administrative units in liberated areas.[29] Numbering some 200,000 soldiers by 1944, the JANL 'despite regular army trappings, remained essentially a guerrilla organization of small, semi-independent, and lightly armed bands' that 'lacked any sort of artillery or formal communications'.[30] It had spent much of its time and energy escaping from a series of large-scale German offensives and was unable to advance into the Serbian heartland of the German occupation until late 1944, by when the JANL was also operating in conjunction with the Soviet armies advancing into eastern Europe.[31] And of course throughout the war the Yugoslav partisans had relied on the Soviet Union and other foreign allies to 'divert the attention' of the enemy's military forces and prevent them being focused on wiping out the expeller resistance in Yugoslavia.

In other countries, however, the resistance to occupation by Nazi Germany did not seek to create liberated areas and regular forces but instead used the first-phase modernisation's strategy of working with national or allied regular armies to expel the invaders. In the Soviet Union there were as many as 150,000 Partisan guerrillas operating behind enemy lines within a year of Germany's 1941 invasion, and later in the war the guerrillas operated – as Clausewitz had envisaged when discussing people's war – in conjunction with their state's regular army and its successful campaigns to expel the invaders.[32] Resistance groups operating in Western Europe instead focused on preparing to fight alongside their states' Anglo-American allies when these allied armies eventually landed in Europe. Thus in 1941 the French Resistance movement was ordered by De Gaulle to stop killing German soldiers, as these killings were producing bloody German reprisals, but was reassured that 'as soon as we are in a position to move to the attack, the orders for which you are waiting will be given'.[33] In the meantime the French Resistance bands would clandestinely build up their military strength in preparation for fighting alongside an invasion by allied armies – and also help with preparations for the eventual invasion by information-gathering and sabotage.[34]

After the Second World War, France became the target of anti-colonial expeller insurgencies, in Vietnam and then Algeria, which succeeded by using a pressure strategy rather than a maoist-model preparation strategy. The later, Algerian insurgency is in one sense a more interesting case than the Vietnamese because it involved the inadvertent use of the pressure strategy by maoist-model insurgents. 'Conventionally presented as just another successful mid-20th century colonial rebellion fought on the Maoist model, this war was in fact shaped by its specific time and its particular antagonists; and it offers an interesting example of the military defeat of a revolutionary movement, which nevertheless gained its objective by political means' – such as by stimulating not only 'foreign criticism' but also the 'electorate's war-weariness'.[35] Although the FLN/ALN movement had begun its insurgency in 1954 with the maoist-model intention of eventually shifting to semistate warfare, the deployment of nearly half a million French troops in Algeria prevented the insurgents from establishing 'liberated' base areas and in fact by the end of 1956 the ALN guerrilla units 'had begun to split into ever smaller fractions as the nationalists realized that they would not defeat the French in a military sense'.[36] Their military situation deteriorated even further when the French implemented the 1959 Challe Plan counterinsurgency campaign, which reduced the ALN to making small-scale sabotage or terrorist attacks.[37] But the insurgents were actually on the verge of winning a complete political victory that would result in full independence and the expulsion of the large European settler population as well as the many French troops and administrators. The political pressure exerted indirectly through a war-weary electorate and foreign criticism had gained an inadvertent strategic victory that the maoist strategy could never have achieved through shifting to semistate warfare.

In contrast, France's defeat by the Vietnamese communists' 1946–54 anti-colonial insurgency had been due to an intentional, not inadvertent, use of the pressure strategy – and had also included a militarily successful maoist-model shift to semistate warfare. From the outset it was clear that the French government would be vulnerable to a pressure strategy, as there was so little political support for the war that the French parliament had prohibited the government from sending French conscripts to fight in Vietnam – thereby leaving the French Far East Expeditionary Corps with too few troops, less than 300,000, to defeat the insurgency.[38] The communists' adoption of a maoist-model strategy produced a premature shift to semistate warfare in 1951 that led to bloody battlefield defeats, but the communist commander, Giap, was aware of the advantages of using a pressure strategy against democratic France (as is discussed in the next section) and that this would be the way he would eventually succeed in expelling the French.[39] His pressure strategy continued to include some semistate warfare, and would produce the famous victory of Dien Bien Phu in 1954 when Giap deployed a 50,000-strong force of regular troops and artillery against a besieged French force of 10,000.[40] But months before the French defeat at Dien Bien Phu the French government had already begun negotiations with non-communist Vietnamese politicians about independence and had also agreed to take part in an international peace conference focused on Vietnam and other Asian conflicts.[41] Furthermore, despite the communists' victory at Dien Bien Phu, this Geneva Conference negotiated a settlement that gave the Vietnamese communists only 'half a loaf', the northern half of Vietnam, and converted the southern half into the non-communist state of South Vietnam. The Vietnamese communists would have to adopt the new strategy of consolidating their rule over the new state of North Vietnam and then making it the foreign ally of a revolutionary communist insurgency in South Vietnam.

That communist insurgency in South Vietnam would lead on to the most famous example of an expeller insurgency using a pressure rather than preparation strategy to expel foreign troops. When the US came to South Vietnam's aid in the mid-1960s, there was a rapid escalation in the number of troops deployed until, like France in Algeria, the US had sent a largely conscript army of some half a million troops overseas to fight a counterinsurgency war. The Vietnamese communists would expel this massive force in a similar way to their expulsion of French forces and to the FLN's victory in Algeria – by using the pressure strategy to defeat their foreign foe politically rather than militarily:

> The Communist high command was all along aware that military victory against the Americans was ruled out; the strategic aim was therefore to make the war so *costly* for the United States that it would *tire* and withdraw. Once the Americans were out of the country, victory over the Saigon government was just a question of time, and probably not much time and effort at that.[42]

Some 20 years later Afghanistan's insurgents would win a similar victory against the world's other superpower, the Soviet Union, and its puppet regime in Kabul. Unlike the Vietnamese communists and Algerian FLN, the Afghan mujaheddin were divided into a number of different political-military movements – and most of them seem to have been premodern insurgent bands motivated by religious and localist, ethnic/tribal opposition to the Soviet intervention. Only Massoud's Tajik-based insurgency in the northeast adopted a maoist-model strategy and perhaps only Hekmatyar's Hezb-i-Islami among the Islamic movements had sufficient ideological content to be classified as a third-phase rather than premodern insurgency.[43] Nonetheless, the more than 150,000 mujaheddin insurgents collectively outnumbered the Soviet troops deployed in Afghanistan, who numbered only some 115,000, because 'the Soviets wished to avoid the "Vietnam syndrome". They wanted the war to remain a low-level, local conflict and to avoid any escalation or direct spillover into neighbouring countries'.[44] The Soviet Union's unwillingness to expand the conflict into neighbouring countries allowed the mujaheddin in southern and eastern Afghanistan to benefit substantially from foreign assistance, as neighbouring Pakistan not only provided a sanctuary and other forms of assistance but also was a conduit for aid coming from other countries, notably the US, that supplied weapons or funds to the insurgents. As early as 1986 the Soviet leadership had decided to withdraw its forces from Afghanistan, and the 1988 Geneva Agreements made this a formal undertaking that was duly carried out by early in the following year. But the mujaheddin's disunity and their weakness in waging semistate warfare (apart from Massoud's forces in the northeast) delayed until 1992 their follow-up destruction of what remained of the Soviet-sponsored puppet state based in Kabul. In contrast to the fall of South Vietnam, this produced only a weak, fragmented state that a few years later would in turn be destroyed and largely replaced by the Taliban's Islamist semistate.

As in the Vietnamese and Algerian cases, the key victory – expelling the foreign state – had not been accomplished by the maoist-model approach of shifting to semistate warfare and winning a war of conquest. Instead in all three cases it was a pressure strategy that had inadvertently or intentionally brought dramatic victories to insurgents fighting against very powerful states. This confirms the impression that the famous maoist-model strategy of preparing for semistate warfare has been overrated as an insurgent strategy and in fact began becoming outdated within a decade or so of Mao's victory in 1949.

An outdated strategy

It can be argued in Mao's own materialist, Marxist terms that technological progress and the increasing urbanisation of Third World countries have certainly made the maoist-model strategy less feasible and in some cases virtually suicidal. The introduction of new counterinsurgency techniques and technology meant that by the 1960s guerrillas were at an increasing disadvantage on the battlefield and when establishing semistate base areas amongst inaccessible terrain. The many revolutionary insurgents who sought to emulate Castro's success in Cuba in 1956–9 seem to have forgotten that their countries and decade offered less favourable opportunities – and indeed the opportunities were becoming less favourable *because of* Castro's success. Two years after that success Guevara reminded revolutionaries that the US and its Latin American allies had learnt the lesson of Cuba and 'will not again be taken by surprise'.[45] Already Latin American armies were undergoing US-assisted training programmes in counterinsurgency techniques and being equipped with helicopters, which would prove to be the key piece of new counterinsurgency technology.[46] By 1968 Debray was acknowledging that this transporting of troops by air precluded guerrillas from setting up fixed bases – and that now no area was really inaccessible.[47] In addition to helicopter technology's ability to transport troops and provide supporting firepower, such sophisticated technology as infrared aerial photography was being applied to counterinsurgency.[48] Guevara's band was tracked down in 1967 by the Bolivian army with the help of infrared aerial reconnaissance that could detect 'even a small band of guerrillas in areas too rugged or remote for effective reconnaissance by normal methods'.[49]

That the changing times required a change in guerrilla strategy was implicitly acknowledged by Debray in his espousal of Guevara's *foco* (nucleus) theory of guerrilla warfare. Debray described the rural guerrilla nucleus as a catalyst for urban insurrection, which implied that the guerrilla band was engaged in a typically terrorist, propaganda or provocation strategy – and he even applied the typically terrorist concept of propaganda by deeds, not speeches, to the band's activities.[50] More conventionally, Taber's *The War of the Flea* presented a model of revolutionary guerrilla warfare that envisaged a rural maoist-model strategy but also included a possible Cuban-style variation in which political victory is won without the need for a shift to full-scale semi-state warfare and its battlefield victories.[51] Presumably Taber had in mind the end-game of Castro's Cuban insurgency, which saw the military dictator Batista flee into exile before his army had been defeated in large-scale battle or had lost the major cities to the advancing columns of the small Rebel Army. Such a political rather than military victory seems to have been produced by a variant of the pressure strategy and against the local rather than a foreign state. And certainly it is a pressure strategy that Taber had in mind when he acknowledged that a political victory can be won by anti-colonial 'urban terrorist' assassination and sabotage, for the withdrawal of the colonial power is achieved by the typical pressure strategy of convincing this foreign state that the colony is not a profitable asset but an economic and political liability.[52]

In the 1970s Laqueur's *Guerrilla* implied that the maoist model of guerrilla warfare was outdated. He argued that 'topographical conditions have lost some of their erstwhile importance' in the modern age and that the technological improvement in security forces' mobility and ability to locate the guerrillas had brought a shift in the overall military balance of power.[53] What is more, the political balance of power had shifted, too, because rapid urbanisation had reduced the political importance of the countryside – not surprisingly, he prophesied that the age of the rural guerrilla 'is drawing to a close'.[54] Although his prophesy seemed to be contradicted by the victory of the Nicaraguan guerrillas in 1979, their reliance on urban support and insurrection meant that their victory was actually an exception that proved the rule.

The effects of urbanisation are a fundamentally materialist, Marxist reason for believing that the maoist model is outdated. As is described in Chapter 8, the increasing prominence of the urban environment has reduced the potential for maoist-model insurgencies because it is an inherently unfavourable environment for their strategy. Even a small and weak state can readily defeat an urban version of the maoist-model strategy because both preparing for and engaging in semistate warfare makes insurgents very vulnerable to encirclement and annihilation by the state's military forces. Unless these forces have been distracted by other threats, such as a rural 'second front' or a foreign ally of the insurgents, it is virtually suicidal to pursue a maoist-model strategy in an urban environment.

The vulnerability of maoist-model urban insurgencies is highlighted by the fate of the most impressive of the 1960s–70s 'urban guerrilla' revolutionary insurgencies – Uruguay's 'Tupamaros'. Their National Liberation Movement numbered several thousand combatant or auxiliary members and operated in a capital city that contained about half of Uruguay's population of three million.[55] The Tupamaros were leftist-populists rather than communists but were doctrinally committed to a maoist-like strategy of first creating a situation of 'power duality' in which the insurgents are seen as a parallel or shadow government and then making a *salto* (leap or escalation) that would require 'greater and higher levels of armed struggle' and the destruction of the state's armed forces.[56] By 1971 a visiting British expert on urban insurgency was using the notion of dual power to describe the Tupamaros' operations and was suggesting that they would have a real chance of success if they attempted a mass insurrection.[57] But in fact their movement was easily destroyed in 1972 when the counterinsurgency problem was handed over to the military, which instituted a very effective crackdown that included mass arrests of suspected Tupamaros. Soon afterwards Guillen published a theory of urban guerrilla warfare that criticised the Tupamaros' attempt to establish a 'state within the state' with secret barracks, arsenals, prisons and 'a heavy logistical infrastructure'; he pointed to their destruction as confirmation that urban guerrillas flourish 'not by establishing fixed urban bases but by living apart and fighting together'.[58]

Other revolutionary insurgencies have established much more obvious urban bases than those of the Tupamaros – in fact they have been embryonic semistates – and experienced more drastic military countermeasures. In the 1970s Turkish revolutionary insurgents set up liberated zones in some shantytowns and 'exercised a crude form of sovereignty in those areas' but were easily eliminated by an army offensive in the early 1980s.[59] A similar fate befell the embryonic semistates set up in urban areas by Islamist insurgents in the first half of the 1990s. Some 'liberated Islamic zones' were established in urban neighbourhoods of Algeria, and the short-lived 'Islamic Republic' set up in a shantytown suburb of Cairo had a population of nearly a million.[60] But the Algerian and Egyptian armies' elimination of these Islamist-ruled pockets revealed yet again the virtually suicidal nature of the maoist-model strategy when it is pursued in an urban environment.

Separatist insurgencies have been no more successful than their revolutionary counterparts in establishing semistates in urban areas. In the Northern Ireland of the early 1970s the Provisional IRA temporarily established what have been described as 'rival mini-states' behind the barricades that the Catholic sectarian minority had erected to keep the British state out of their No-Go neighbourhoods. 'Not only did the Provisionals deal out their own forms of "justice" – ranging from tarring-and-feathering to knee-capping and murder – but they also collected protection money, allocated houses on the Housing Executive estates, ran the publicly-provided sports and social centres, and licensed the fleets of black taxis which had replaced the Corporation buses burned on the barricades'.[61] But the British army's massive Operation Motorman in July 1972 reoccupied the No-Go areas with little difficulty or bloodshed.[62] The Provisionals had been forewarned and had decided to escape across the Irish border or to merge back into their local neighbourhood communities, where they would exercise a political rather than territorial form of dual power/sovereignty.

Expeller insurgencies face the same problems in protecting embryonic urban semistates and also provide some famous examples of the futility of shifting to semistate warfare in an urban environment. The need for rural or foreign allies was illustrated by the Polish Home Army's 1944 uprising in Warsaw against the German occupation. The uprising was attempted only because advancing Soviet armies had inflicted heavy defeats on German forces and were approaching the city but, for whatever reasons, the Home Army in fact had to fight without Soviet assistance. Although this underground army numbered tens of thousands and took over large parts of the city, it was eventually defeated by a

40,000-strong German force equipped with tanks, artillery and air support – which methodically retook the city and inflicted hundreds of thousands of casualties on the combatant and non-combatant population.[63] Similarly, the Vietnamese communists' bloodiest military defeat, costing them some 45,000 casualties, was the urban-oriented 1968 Tet Offensive; it gave the US and South Vietnamese forces the opportunity to use their superior firepower – including artillery and air strikes – against a more accessible and concentrated foe than they had been fighting in rural areas.[64]

More recently, the Iraqi insurgents who had turned Fallujah into 'an Islamic mini-state' and 'an insurgent sanctuary' had no answer to the heavy weapons and superior training of US forces when they retook the city in November 2004.[65] In fact many of the insurgents seem to have realised the futility of fighting in these circumstances and about half of them had fled the city before the offensive began, which was one reason why in November and December insurgent violence 'actually spread far beyond the city to other relatively quiet areas and cities'.[66] It was an object lesson in the advantages that urban insurgents gain from resisting the territorial, semistate temptation and instead following the third phase's approach of staying vertical and using pressure, propaganda or provocation strategies.

Hezbollah and Hamas seem to be two prominent exceptions to this tendency, as each has established a semistate and has recently fought an expeller war against the state of Israel, respectively in 2006 and 2009. But both are exceptions that prove the rule. They are typical third-phase insurgencies in having religious-based ideologies (respectively derived from the Shiite Islamic revolution in Iran and from the Sunni Muslim Brotherhood founded in Egypt in the 1920s) and in preferring not to make a maoist-model shift to semistate warfare against their enemy state.[67] Both had established their semistate rule in, respectively, southern Lebanon in the late 1980s and the Gaza Strip in the early 2000s as the *result* of a successful expeller insurgency, not as preparation for shifting to semistate warfare. And despite both Hezbollah's and Hamas's use of unguided rockets as a 'poor man's V2', there is no prospect of either semistate going further and attempting to use regular armies to destroy the state of Israel. Even the 2006 and 2009 ground fighting was a matter of defending against temporary Israeli incursions and would have resulted in military suicide if these semistates had sought victory through battle, not political pressure.

Pressuring politically through military hurt

The pressure strategy may have been used in practice even in ancient times but it was not made explicit and put in writing until Mao's 1938 article 'On Protracted War' pointed to the political problems facing the Japanese invaders on their home front.

> Signs of exhaustion are beginning to appear in his [the Japanese state's] finances and economy; war-weariness is beginning to set in among his people and troops; and within the clique at the helm of the war, 'war frustrations' are beginning to manifest themselves and pessimism about the prospects of the war is growing.[68]

These problems would prevent the Japanese government from increasing the war effort in China and indeed might lead to a Japanese withdrawal from China, especially if war-weariness led to a political revolution – as it had in war-weary Russia in 1917 and war-weary Germany in 1918.[69]

Mao's argument was taken a step further in the early 1950s by the Vietnamese communists' military leader, General Giap, in a speech he made to the political commissars of troops fighting to expel French colonial rule. He recognised that France's political leaders, unlike Japan's in 1938, were subject to the institutional and popular pressures of a democratic political system. Giap predicted that 'public opinion in the democracy will demand an end to the "useless bloodshed," or its legislature will insist on knowing for how long it will have to vote astronomical credits without a clear-cut victory in sight'.[70] Such political pressure upon the government 'forces the democratic politicians to agree to almost any kind of humiliating compromise rather than to accept the idea of a semi-permanent antiguerrilla operation'.[71]

The pressure strategy exerts political pressure *indirectly*, by stimulating the enemy state's legislature, electorate, public opinion and other politically influential entities to exert political pressure on the government. This indirectness is an implicit aspect of Schelling's analysis of why the military power to hurt is politically effective, as the people being hurt clearly having some influence over the state's decision-makers.[72] On a state-to-state basis this political use of the military power to hurt was famously labelled by Schelling as the 'diplomacy of violence' or 'coercive diplomacy'. But he also noted that it may occur in such an extreme context as the First World War's long and bloody battle of Verdun; the German strategy of killing hundreds of thousands of French soldiers in a battle of attrition may well have been aimed at making the loss of 'sons, husbands, fathers and the pride of French manhood' so hurtful to French society 'as to be unendurable' and 'to make surrender a welcome relief'.[73] Insurgents may be aiming to produce the same sort of effect by killing the soldiers of a foreign state or their local state in sufficient numbers to make the war unendurable to the foreign society or to the sector of local society that is opposing the insurgents' revolutionary or separatist political goals.

It is this ability to inflict mental hurt on a society by inflicting physical hurt on its soldiers which enables the pressure strategy to be used without resorting to a terrorist mode of operations. There is no need to resort to targeting civilians if they will pressure their political leaders to end the mental hurt they are suffering because of the physical hurt inflicted on their soldiers. (Another way of indirectly inflicting hurt on the civilian population, as was noted by Giap, is through the economic hurt of the financial costs of the war.) Of course the military power to hurt has often been used by insurgents and states in a terroristic manner, as is described in Chapter 4, but the present chapter is concerned with the military power to hurt being used indirectly, not directly, against civilians.

Irrespective of whether insurgents use the pressure strategy in a terroristic manner, its effectiveness varies markedly according to the type of insurgency – and it is a particularly ineffective strategy for achieving revolutionary political goals. Perhaps the nearest it came to success was in the early 1970s in Argentina, when the Peronist and Marxist insurgents used a pressure strategy to push the country's military government into carrying out a democratisation. But this was a very unusual case of insurgents seeking to bring about revolution indirectly – through a democratisation that would in turn lead to an undemocratic revolution.[74] For the insurgents expected that free elections would result in the return to power of Juan Peron, a populist President of the 1940s–50s, who would then use his position as President to bring about a socialist revolution 'from above'. An assortment of Peronist and Marxist insurgent bands therefore sought to increase the political pressure on the military government to democratise – pressure coming from within the military as well as from civilian society – by making the country appear virtually ungovernable. Assisted by increasing trade-union militancy and serious economic problems, the insurgents increased the pressure on the government by 'expropriations of money and arms, kidnappings of representatives of foreign companies, dispensation of revolutionary justice [assassinations], Robin Hood-style actions to help the poor, and bold moves to reveal the government's weaknesses'.[75] The military government eventually gave in to the pressure and held democratic elections in 1973, which duly resulted in Peron becoming President.

However, the terminally ill Peron would prove to be a democratic centrist rather than a revolutionary socialist, and after his death some of the revolutionary bands returned to the more standard 'urban guerrilla' strategies. This time the military responded by returning to power and carrying out a very repressive and violent 'dirty war' that destroyed the revolutionary insurgent bands. So the only situation in which a pressure strategy seems likely to bring revolutionaries some degree of success is when they are willing to accept the partial success of democratisation and peaceful participation in democracy, as occurred in the cases of El Salvador, Guatemala and Nepal that were mentioned earlier – and which seem to have occurred through an *inadvertent* use of a pressure strategy by maoist-model insurgents.

The separatist insurgencies seem to have been more successful in their use of the pressure strategy. In particular, the IRA's implicit or inadvertent use of a pressure strategy against the British government in 1919–21 won a famous victory that may have influenced Mao's 1938 account of the pressure strategy as much as did the First World War examples of the political effects of war-weariness. However, the historical record since the Irish success seems to indicate that the best any separatists can hope for from the pressure strategy is the partial success of a compromise negotiated settlement – like that achieved in the 1990s by the Provisional IRA's insurgency in Northern Ireland. The task facing the Provisional IRA had been foreshadowed in the early 1970s by two remarks made by British Cabinet Ministers; one had spoken of 'achieving an acceptable level of violence' in Northern Ireland and another had pointed out in secret talks with the PIRA that the government could 'accept the casualties' as 'we probably lose as many soldiers in accidents in Germany' in the large British army then stationed in West Germany.[76] Such a self-confident attitude would have been much less likely if the British government had been dealing with an expeller insurgency in a colony rather than a separatist insurgency in a region that was regarded by international and domestic public opinion to be a part of Britain.

Indeed it is the expeller insurgencies that have found the pressure strategy to be very effective and capable of achieving not merely partial but complete success. The reasons why it has been so successful were explained in the mid-1970s by the most famous Western version of

Mao's and Giap's insights about the pressure strategy. Mack's article, 'Why Big Nations Lose Small Wars', described the defeat of a big foreign state by relatively weak local insurgents as a case of destroying a state's political capability or 'will' to continue a war.

> In every case, success for the insurgents arose not from a military victory on the ground – though military victories may have been a contributory cause – but rather from the progressive attrition of their opponents' *political* capability to wage war. In such asymmetric conflicts, insurgents may gain political victory from a situation of military stalemate *or even defeat*. ... Lacking the technological capacity or the basic resources to destroy the external enemy's military capability, they must of necessity aim to destroy his political capability. If the external power's 'will' to continue the struggle is destroyed, then its military capability – no matter how powerful – is totally irrelevant.[77]

In a sense Mack was redeploying Clausewitz's classic cost-benefit analysis of limited war between states. The latter had argued that one of the grounds for making peace is the 'unacceptable cost' of victory – unacceptable in terms of the political object or goal of the war.[78] In the logical fashion of a syllogism he argued that since war 'is controlled by its political object, the value of this object must determine the sacrifices to be made for it in *magnitude* and also in *duration*. Once the expenditure of effort [cost] exceeds the value [benefit] of the political object, the object must be renounced and peace must follow'.[79]

But Mack introduced a new factor by highlighting what he termed the 'asymmetric' nature of the two sides' political goals. While the expeller insurgents' political goal is so valuable that they will continue fighting until they have lost their military capability, the foreign state has a much less valuable political goal and certainly is not fighting for its survival as an independent state. In these asymmetric circumstances there is an obvious political temptation for the foreign state to withdraw from a costly and protracted foreign war against insurgents who clearly lack the military capability to threaten the state's survival.[80] In domestic political terms, too, there is 'the potential for domestic conflict in the homeland', even among the ruling elite of a non-democratic political system, and these political divisions within the elite and/or public about whether to continue the war 'become *in themselves* one of the political costs of the war', just as the economic resources spent on the war, and any war-related taxation or inflation, are an economic cost inflicted on the society.[81]

The political calculus of the pressure strategy

The insurgents' pressure strategy is based upon the cost-benefit political calculus used by the enemy state's careerist political leaders. Mack pointed out that when the foreign state's political leaders decide to end the war, it is 'because they have no choice' rather than because they have had 'a sudden change of heart'.[82] Unless perhaps they are absolutist rulers, the leaders of dictatorships as well as democracies use a political calculus that assesses the costs and benefits to their political careers of making such policies as when to end a costly counterinsurgency war fought for limited goals. It is true that leaders who have started wars often risk their careers (in terms of losing office or the verdict of history) rather than implicitly acknowledge that the war is no longer worth fighting, if it ever were, and that it is time for both the nation and its leaders to cut their losses. However, sooner or later the war-starting leaders or their successors can be expected to take a realistic, rational view of their careerist self-interest and follow the dictates of the political calculus.

Democracies' tendency to show a weaker political will/capability than dictatorships to oppose an insurgent pressure strategy can be readily explained by the nature of their political system and of their leaders' political calculus. There is greater freedom of not only information but also political competition in democratic political systems, and it includes electoral competition for the state's leadership positions as well as the muted competition among policy advocates and public administrators. Thanks to this freedom of information and competition, a counterinsurgency war's failings will eventually become public knowledge and will eventually elicit an appropriate adaptive response from politically calculating policy makers.[83]

The political calculus's assessment of the costs and benefits involved in continuing/ending the counterinsurgency war will usually include a number of different factors, multipliers, relationships and indirect effects. For example, the leaders of a foreign state that is fighting an expeller insurgency usually justify the war in terms of the benefit of protecting important national interests, and Mack argued that the greater the salience of these interests, the greater the resistance in the foreign state to any proposals for withdrawal from the insurgents' territory.[84] But there is room for debate about the amount of benefit to be gained from protecting these interests because clearly they are not vital interests or else this would not be such a limited war. Mack described how even a dictatorship can be plagued by controversies within its ruling elite about 'the costs and benefits' of continuing with a war that involves important but not vital interests.[85] And even a dictatorship may make an assessment of costs and benefits that takes a broad view and establishes priorities, as when the Soviet Union decided in 1986 that the counterinsurgency war in Afghanistan was having too costly an effect upon the new foreign-policy priority of improving relations with the West.

The complexity of the political calculus is greater in democracies, where there is a wider range of political pressures on leaders who are faced with the problems of priorities within and between policy areas. A famous example is the situation President Johnson faced in March 1968 in the aftermath of the Vietnamese communists' Tet Offensive and of the resulting request from his military commanders for 206,000 more troops to be sent to Vietnam. Johnson was well aware that 'press and popular opposition to the war was mounting, and many prominent businessmen were worrying openly about the damage that the war-generated government deficits and drain on American gold reserves were having on the national economy' but a more immediate problem was Congress's unwillingness to fund the new troop requests:

> What concerned him most was how the \$5–7 billion of additional war costs associated with Westmoreland's requests would take resources from his Great Society programs. [He noted privately that] 'our financial situation is abominable' and that Congress would cut spending on programs for 'poverty, housing and education' before it would raise taxes and increase war spending.[86]

The situation is reminiscent of Giap's earlier-quoted prediction that a democracy's 'legislature will insist on knowing for how long it will have to vote astronomical credits without a clear-cut victory in sight' against an expeller insurgency. However, President Johnson was facing a broader and more complex policy and political situation; it was at the end of this month that – apparently hoping to salvage his reputation and place in history – he announced that he would not stand for re-election.[87]

The complexity of the political calculus also offers insurgents a range of opportunities for indirectly exerting pressure on the enemy state's leaders. For example, in some circumstances pressure may be exerted more effectively and efficiently through reducing benefits rather than by increasing costs – through targeting the enemy state's goals rather than its soldiers. Clausewitz pointed out that one of the grounds for making peace in a limited war is simply the futility factor, or what he termed the 'improbability of victory', even though the futility factor has its biggest impact in any form of warfare when it leads to a unit, army or nation deciding to surrender rather than continue a futile resistance.[88] This factor can also be exploited by insurgents fighting to expel an invading or occupying foreign state, as was illustrated by the Israeli invasion and occupation of southern Lebanon in the 1980s. The most devastating attack on the Israeli occupation was the 14 September 1982 bomb that killed Israel's key local ally, President-elect Bashir Gemayel: 'As a result, Israel lost control over events in Lebanon'.[89] The occupation's ultimate, political goals had become unattainable and there would be growing debate and doubt among Israelis about the benefits of an occupation that was finally ended in 1985 by the Israeli withdrawal from most of southern Lebanon.[90]

Similarly, insurgent attacks against the US occupation of Iraq were aimed at not only inflicting costs but also preventing the US from achieving its political and economic goals.[91] The most important example was the sectarian attacks on Shiites that began in 2004. These attacks on Shiites 'were designed to goad them into a counterresponse, one that would increase the level of violence immeasurably, render the US presence untenable and thwart the elections'.[92] The goading attacks continued after the 2005 elections and became identified with Al Qaeda in Iraq (AQI) under Zarqawi's leadership in 2005–6, but they soon became an example of the unintended consequences that so often result from complex schemes. First, the Shiite response to the sectarian goading took the form of a tacit alliance between Shiite-dominated governments and the Shiite militia bands to introduce the classic counterinsurgency strategy of 'resettling' the Sunni population of Baghdad and other politically sensitive areas (though the resettling sometimes seems to have been a sectarian version of ethnic cleansing) so as to remove the social support base for AQI and other Sunni insurgent bands.[93] Second, the resulting increase in inter-sectarian violence led the US in 2007 to introduce in Sunni-inhabited areas the other classic counterinsurgency strategy of deploying troops in the villages or neighbourhoods and seeking to create alliances at this local level that would produce local militias capable of helping the troops to control the area and counter the insurgents.[94] The combination of these two counterinsurgency strategies not only eventually led to a marked reduction in the level of violence but also markedly reduced the presence and effectiveness of AQI in Baghdad and much of the Sunni-inhabited region of Iraq.

Although the AQI's pressure strategy was successfully countered, it seems to have identified the most effective and efficient way of increasing the political pressure on the US government to withdraw its forces from Iraq. A study of public support for President Bush during the 2003–4 political campaigning season had indicated that the public's casualty 'sensitivity' or 'tolerance' was linked to their views about whether it had been right to invade Iraq and to their 'expectations of success' in achieving the war's potential benefits – and 'success appears to be the more important factor'.[95] In turn this was success in achieving the potential benefits of a) Iraqis enjoying stable and democratic government, b) Iraqis providing for their own security and c) Iraqis living peaceful and normal everyday lives.[96] If the AQI insurgents had succeeded in provoking a sectarian civil war, these potential benefits would have seemed virtually unattainable. And then the futility factor would presumably have had a marked effect on the public's attitude, even if there had been no significant increase in the number of casualties suffered by US forces in Iraq. In that sense the futility factor would be operating as a hurt-multiplier that increases the sensitivity to casualties and therefore the amount of hurt that the insurgents engender by inflicting casualties. It is an increase in the mental cost of the war that is felt by the public and indirectly by the political leaders who take the public's attitude into account as part of their political calculus of the careerist costs and benefits of continuing the war.

The futility factor is not the only form of hurt-multiplier that increases the costs of fighting against expeller insurgents. In recent times some societies have experienced long-term cultural changes that seem to have increased the sensitivity to casualties. Luttwak suggested in the 1990s that one of the reasons why the US should adopt a post-heroic, casualty-minimising attitude to warfare was the unwillingness of post-industrial societies with low birth rates to suffer casualties in avoidable wars.[97] Whatever the reason for this cultural change in sensitivity to casualties, Luttwak's prognosis was apparently confirmed a decade later by Mueller's analysis of the American public's markedly reduced tolerance for casualties since the days of the Korean and Vietnam wars. He pointed out that the public had become disenchanted with the war in Iraq after only 1,500 combat deaths, as compared to the 20,000 combat deaths that had been required before the public of the 1960s became disenchanted with the war in Vietnam.[98] (Furthermore, there was no prospect of the Iraq conflict leading to conscription and the spread of casualties to parts of society that were unlikely to be attracted by the notion of serving their country in a dangerous and poorly paid manner.) Although this comparison did not sufficiently take into account the benefit/success/futility factor, it indicates that Western societies have become more emotionally or morally concerned about servicemen and their families suffering hurt on behalf of the state when its survival or vital interests are not at stake.

Furthermore, some members of these Western societies have apparently become intolerant of hurt suffered even by their enemies. Merom has argued that when democracies are fighting a counterinsurgency war in a foreign land, they are limited by a varying 'political capacity' that includes their societies' intolerance of 'brutal engagement of the enemy'.[99] The problem of low levels of tolerance for brutally engaging the enemy is a relatively new aspect of states' political vulnerability, and apparently has been produced by the rise of democracy and of a politically influential sector of society imbued with liberal values:

> Traditionally, the use of violence abroad did not involve difficulties at home. Subjects were often unwilling to sacrifice their money or life to underwrite their leaders' military adventures abroad, but they did not care about the fate of foreigners, be they insurgents or the civil population that supported the latter. ... Ultimately, only the development of democratic political institutions and an educated liberal constituency in the West have changed this state of affairs.[100]

Some recent confirmation of this intolerance of brutality can be found in the marked drop in the American public's support for the Iraq war following the revelation that detainees had been abused at Abu Ghraib prison.[101] But a better example is described in Chapter 4, when it shows how the Algerian FLN's actual provoking of politically counterproductive repressive measures, notably the use of torture, resulted in a further erosion of the French public's support for the war and more political pressure on their leaders to withdraw from Algeria.

In fact by merely prolonging the war an insurgent band can inflict political costs upon the foreign state's leaders directly as well as through the political pressure exerted on them by their fellow citizens. Mack pointed out that time is a limited political resource and that the amount of time spent debating a controversial asymmetric war may be 'to the detriment of the pursuit of other objectives'.[102] This political 'opportunity cost' is experienced by the leaders of any state, whether democratic or non-democratic, who find that a protracted war continually consumes time that they would have preferred to spend on other policy matters. In addition to this opportunity cost, a prolonged counterinsurgency war can cost the state's leaders some of their reputation for competence, as the failure to win a quick victory against a much weaker foe gives good grounds for political competitors and the public to question the leaders' competence, especially if they have raised expectations of an early victory.[103]

Finally, a democracy involved in any prolonged conflict faces the danger of political attrition. This political cost was identified by the Provisional IRA in the early 1980s and incorporated in their 'long war' pressure strategy. The PIRA believed that 'Eventually, dissensions over "shoot-to-kill" allegations, hunger strikes, issues of civil liberties, law and order, etc. would adequately muddy the waters concerning the legitimacy of British involvement in Northern Ireland to destroy domestic consensus, and so undercut Britain's political will to stay'.[104] In this and so many other cases of insurgent stateless warfare, the means by which war is fought are almost as political as the end for which the war is fought. More than 40 years ago Galula pointed out that 'so intricate is the interplay between the political and the military actions that they cannot be tidily separated; on the contrary, every military move has to be weighed with regard to its political effects, and vice versa'.[105] The assessment of the political effects is most needed – but is so often lacking – when the military move is directed against a civilian target in the terroristic manner discussed in the next chapter.

4
Terroristic Warfare and Strategies

When an insurgent strategy includes the attacking of civilian targets, it focuses attention on another aspect of how insurgents use organised violence – namely, what mode of operation is being used and, particularly, whether it is the terrorist mode. As was evident in Chapter 2's discussion of mode, whether to categorise an attack on a civilian target as 'terrorist' is often a matter of interpretation or context. And a further complication is that there are obvious similarities between the terrorist mode of insurgency and the civilian-targeting that has occurred in warfare between states. Studies of modern total war, notably the two world wars, have highlighted the extent to which states have engaged in this 'terroristic' warfare, particularly when engaged in the strategic bombing of the enemy state's homeland or 'home front'. Therefore the first half of this chapter considers the obvious similarities – and crucial differences – between a) the terrorist mode of operation that insurgents sometimes adopt when pursuing a pressure strategy and b) the terroristic warfare that sometimes occurs when states are pursuing pressure strategies against other states.

The second part of the chapter focuses on the two insurgent strategies, propaganda and provocation, that are typically terrorist and have no apparent equivalents in warfare between states – certainly not in the sense of typically involving terroristic warfare. Finally and as a contrast with these two strategies, there is a brief description of how insurgents have sometimes adopted a terrorist mode of operations when pursuing a preparation strategy but have used this violence against civilians in a restrained, selective manner and only as a supplement to their primary, guerrilla mode of operations. In addition to providing a contrast with the typically terrorist strategies, this brief return to examining the maoist-model preparation strategy that was discussed in the previous chapter is an appropriate ending to these two chapters' extended study of insurgent strategies.

Total war and attacking the civilian 'home front'

In the previous chapter the pressure strategy was described in terms of insurgents increasing the political pressure on the enemy state by indirectly – through attacks on its soldiers – inflicting hurt on its civilian society. But of course sometimes the strategy has taken the terrorist form of directly attacking the enemy state's civilians. Using the pressure strategy in this terrorist mode is comparable to states' use of strategies that involve total war and its terroristic manner of attacking an enemy state. In fact more than 20 years ago Jenkins suggested that insurgent terrorism was drawing on the conceptual legacy of total war between big nations:

> Modern terrorism also draws on a legacy of World War II; the concept of total war, in which virtually ... every aspect of human production became a legitimate target. Modern terrorists extended this concept. The planners of the strategic bombing campaigns of World War II recognized in theory the existence of civilian bystanders; in practice of course the violence was indiscriminate. Today's terrorists recognize few innocent bystanders either in theory or in practice.[1]

That states' strategic-bombing campaigns recognised the existence of 'innocent civilians' – whether they be women, children, babies or the elderly – certainly did not stop them from being killed by the thousands, tens of thousands or even hundreds of thousands in bombing attacks on homelands that had become 'home fronts' and as much a target as the military fronts that they supplied with their industrial production. In recent times the evaluation of these Second World War bombing campaigns has viewed their 'area' bombing of cities (as distinct from their attempted 'precision' bombing of industrial targets) as more of a deliberate than an unintended killing of civilians and as the equivalent of terrorism by insurgents. For example, a philosopher who has studied this strategic bombing has recently argued that 'in one crucial respect – one respect only – there is a dismaying similarity between area bombing and terrorist bombings: namely, that they both seek to coerce a people by blowing up as many of them as possible and thereby terrorising and demoralising the rest'.[2] In fact if both are seeking to coerce a people by terrorising them, then both are engaging in the pressure strategy of indirectly exerting political pressure on the enemy state by deliberately inflicting hurt and fear on its ordinary civilians; the insurgents are resorting to a terrorist mode of pressure strategy and the area-bombing state is resorting to a terroristic warfare aimed at political rather than military victory.

The same sort of issues about the economic or political aims of attacking the home front had arisen with the use of naval blockades. As Schelling pointed out, the use of the blockade weapon had involved a crucial ambiguity that would recur when bombardment weapons were used to attack enemy territory: 'It has never been quite clear whether blockade – of the South in the Civil War or of the Central Powers in both world wars, or submarine warfare against Britain – was expected to make war unendurable for people or just weaken the enemy forces by denying economic support'.[3] The same ambiguity would recur in the Second World War when bombing enemy homelands was rationalised either a) as showing civilians that 'surrender was better than persistence' or b) as weakening the enemy's war economy, whether in a general way or by selectively denying key supplies to the enemy's military forces.[4]

A further ambiguity arose from *indiscriminate* attacks on economic targets, in which civilians were hurt as part of the 'collateral damage' inflicted through imprecise bombing of economic targets. For example, in 1940 both the German and British air forces found it impossible to attack 'single military targets with existing air technology without spreading destruction over a wide circle around them. This explains why both sides believed the other was conducting a terror campaign against civilian morale' when in fact they were carrying out indiscriminate attacks on military and economic targets.[5]

These ambiguities and issues had been foreshadowed in the earlier world war by Germany's use of Zeppelin airships to drop bombs on Britain. The Zeppelins' bombing of London was initially restricted to military targets, such as the docks, but soon virtually all restrictions were lifted by the government and the key target became the City financial area.[6] Although this was an economic target, there was also some intent to create a panic that would bring an end to the war, even though the Zeppelins dropped in total less than 200 tons of bombs – killing 557 people – and created much more outrage against the 'baby killers' than panic among the Londoners.[7]

However, the first air campaign against civilian targets was perhaps less important than a post-war theoretical exposition that has become a classic of military strategy: Douhet's 1921 *The Command of the Air*. How influential this book was at the time is debatable but it certainly epitomised a very influential trend of thought about the use of air power. Douhet argued that air power would dominate future wars and bring quick victories for the dominant air powers, for air attacks on the enemy's home front would be aimed not only at industrial targets but also at the civilian population – which, 'driven to the point of madness, would force the government to surrender'.[8] This seems to be the ultimate and most explicit expression of a pressure strategy based upon directly and deliberately hurting ordinary civilians.

Such an approach was not confined to an isolated or obscure military theorist. In the 1920s, too, the commander of the British air force, Trenchard, was arguing a similar thesis that emphasised the importance of attacking civilian 'morale'. He even predicted that 'the French in a bombing duel would squeal before we did', though later in the decade he conceded that it was inhumane to bomb a city 'for the sole purpose of terrorising the civilian population' rather than for the military objective of demoralising workers in munitions industries.[9]

By the 1930s the term 'terror bombing' was being used to describe Douhet-like attacks on civilian morale.[10] When war broke out in 1939 between Germany and Britain, though, both sides had sufficient bombing forces 'to deter the other from using them for terror bombing attacks' and Prime Minister Chamberlain made a principle out of necessity by declaring that Britain would never deliberately attack civilians 'for the purpose of mere terrorism'.[11] As for Hitler, he was not prepared to authorise terror bombing against Britain, even in September 1940 when he shifted the axis of the Battle of Britain to attacks on London and other urban areas. He authorised 'a general campaign against urban targets and enemy morale' but 'rejected the idea of inducing "mass panic" through deliberate attacks on civilian areas'; instead the 'political will to resist was to be broken by the collapse of the material infrastructure, the weapons industry, and stocks of fuel and food'.[12] Nonetheless, his air force's lack of bombing accuracy meant that in practice these raids were indiscriminate attacks that would kill some 40,000 civilians in 1940–1 and would appear to the British to be a Douhet-like attempt to panic them into surrender.[13]

Not surprisingly, the British retaliated with a bombing campaign directed at German cities. In July 1941 RAF Bomber Command was given a directive that included destroying the morale of Germany's civilian population and particularly the morale of its industrial workers, and in February 1942 it was given the directive that this destruction of morale was to be the primary objective.[14] Of course the emphasis on industrial workers' morale provided an industrial-target justification for night-bombing's inaccurate area-bombing of cities, which often included dropping large numbers of incendiaries in what became known as 'fire-bombing'.

The British retaliation in turn produced a more extreme approach by the enemy but this was also a product of desperation, as Nazi Germany was facing the increasingly likely prospect of defeat. When Hitler began bombarding London in 1944 with his V-weapon cruise and ballistic missiles, he was hoping that their indiscriminate explosions would devastate London and lead to the overthrow of Churchill's government.[15] As Schelling pointed out, 'Hitler's V-1 buzz bomb and his V-2 rocket are fairly pure cases of weapons whose purpose was to intimidate, to hurt Britain itself rather than Allied military forces' and 'to get the war stopped before suffering military defeat'.[16] They were also fairly pure cases of aerial bombardment aimed at creating a Douhet-like political pressure for ending the war.

The other fairly pure case of a Douhet-like pressure strategy that occurred in the Second World War was the US strategic bombing of Japan. Even before the August 1945 use of nuclear weapons, the US bombing of Japan had been changed by General LeMay into a night-bombing campaign of firebombing Tokyo and other major cities that killed some 230,000 civilians.[17] LeMay justified this killing of civilians by employing a new form of the industrial-targeting argument. 'There are no innocent civilians' because the cottage-industry basis of Japanese production meant that the 'entire population got into the act and worked to make both aeroplanes or munitions'.[18] But it is more difficult to apply any industrial-target argument to the nuclear attacks that culminated LeMay's campaign and also foreshadowed a new era of civilian-targeting. Schelling argues that the use of atomic bombs against Hiroshima and Nagasaki was 'not mainly an attack on Japan's material strength'; these 'weapons of terror and shock' were a threat of more devastating hurt to come and were aimed mainly at a political target rather than the factories and people that they destroyed.[19] That political target was the increased political pressure which would be exerted on the Japanese government to end the war by an unconditional surrender.

By the 1960s it seemed that such politically pressuring civilian-targeting was occurring in both smallest-scale and largest-scale warfare – at both the lowest and highest intensity of military activity. At one end of the scale and intensity was the nuclear age's capability of hurting civilians much more extensively and quickly than with non-nuclear bombardments.[20] At the other end of the scale and intensity there was what Schelling described as the 'brutal war' of insurgency and counterinsurgency that had recently occurred in Algeria, which 'was more a contest in pure violence than military strength; the question was who would first find the pain and degradation unendurable'.[21] In this new era:

> noncombatants appeared to be primary targets at both ends of the scale of warfare; thermonuclear warfare threatened to be a contest in the destruction of cities and populations; and, at the other end of the scale, insurgency is almost entirely terroristic. We live in an era of dirty war.[22]

Similarities with the terrorist mode of insurgency

There are other important similarities between an insurgent band's terrorist-mode pressure strategy and a state's terroristic pressure strategy. Insurgents' vertical attacks by an enemy who is 'everywhere and nowhere' have similar advantages to states' vertical means of attacking one another's civilians by bombarding them from the air or aerospace. Schelling emphasised that this bombardment is a means of inflicting 'coercive pain and suffering directly' on the enemy nation and of 'getting at' the enemy's civilians without first having to defeat the military obstacles in the way; by the 1940s strategic bombers and V1/V2 missiles were flying over these obstacles and attacking the enemy's territory directly.[23] In comparison, insurgents have throughout history been able to 'get at' the enemy state's civilians and/or attack its territory without first having to defeat the military obstacles in the way. They can pursue a pressure strategy that hurts the enemy state's civilians directly, not merely by the indirect means of hurting their soldiers, and in fact this direct hurting can take the form of attacks that deliberately target ordinary civilians. As the deliberate targeting of ordinary civilians is the ultimate expression of insurgency's terrorist mode, it is not surprising that such insurgents are commonly referred to as 'terrorists' rather than 'guerrillas' or even 'insurgents'.

But the similarity between insurgents' and states' ability to inflict hurt directly on civilians has also extended to the ambiguities and complexities involved in states' bombardment of each other's home front. It is true that one source of ambiguity and complexity is actually removed by insurgents' obvious inability to wreak significant destruction on an enemy state's economic or industrial capacity to supply its military forces. Insurgents' attacks on economic targets are instead aimed at inflicting sufficient economic hurt on civilians to increase the political pressure on the enemy state. However, this is similar to the ambiguity involved in aerial bombing that is aimed not at creating a Douhet-like panic but at other forms of hurting civilians that would make the war politically unendurable, such as Hitler's decision in September 1940 to break Britain's political will by a bombing campaign that would not resort to terror bombing.

Furthermore, insurgents' indiscriminate attacks on civilian targets suffer from a similar ambiguity to the indiscriminate bombing of civilian targets that results from an inherent lack of accuracy. As was noted earlier, the German bombing campaign of late 1940 was in fact interpreted by the British as a terror-bombing campaign because it was killing thousands of civilians through its inaccuracy, which was an inherent lack of accuracy in that era of bombing technology. The same situation occurs when insurgent bombs are inherently lacking in accuracy and are likely to kill ordinary civilians as well as the economic or political target of the bomb. In fact indiscriminate attacks that are inherently lacking in accuracy always raise the question of whether the inaccuracy is being used as a convenient excuse for something more extreme and terroristic, just as inherently inaccurate aerial bombing directed at industrial targets can be used as an excuse for the killing of ordinary civilians through the inevitably massive collateral damage.

Yet perhaps the crucial similarity between the terrorist mode of insurgency and states' terroristic manner of warfare is that both have so seldom succeeded in pressuring their enemy into ending the war. Of course these states were fighting total wars of survival, not limited wars, and so it is hardly surprising that they refused to be pressured by the pre-nuclear levels of hurt inflicted by bombers and missiles. Similarly, revolutionary insurgents should not expect to destroy and replace their local state by using some sort of pressure strategy. As Mao recognised long ago, revolutionary insurgency is a 'war of survival' for the state – it can be defeated only by militarily destroying its forces and taking its territory.

Expeller insurgents have more reason than revolutionaries to expect a terrorist mode of pressure strategy to be successful. Their foreign foe is fighting a limited war, not a total war of survival, against the insurgents and therefore would be more willing to end the war in order to end the terrorist attacks on its homeland. However, the expeller insurgents are faced with the problem that they can inflict so little hurt on the enemy's homeland that the attacks' only political effect might be to produce public outrage and hostility towards the insurgents and thus actually reduce any political pressure on the foreign state's leaders to end the war. After all, public outrage and hostility was the main effect of Germany's First World War airship bombing campaign against Britain, which involved more attacks and tonnage of bombs than an expeller insurgent band could expect to deliver against a foreign state's homeland.

So it is not surprising that the three most famous cases of expeller pressure strategies did *not* include attacks on the foreign state's homeland. The Afghan mujaheddin and the Vietnamese communists did not attempt to attack the Soviet, the US or French homelands, and the Vietnamese communist leaders would not even allow attacks on US civilians working in South Vietnam – a series of such attacks occurred in 1964 but was a local initiative that was quickly terminated by higher authority.[24] The exception that proves the rule is that the Algerian FLN's inadvertent pressure strategy against France eventually did produce a short-lived pressure campaign of bombings in the French homeland.[25] But the attacks in August–October 1958 sought to avoid civilian casualties and were aimed at military/police or economic targets, such as an oil refinery near Marseilles, petrol storage tanks in Rouen, and a munitions factory and police garage in the Paris area.[26] Nonetheless, these attacks 'designed to "bring the war home to France" and disrupt life in the metropole aroused negative responses' and it was this 'backlash of hostility that apparently dissuaded the FLN from further attempts of this nature' – as the FLN leaders 'now understood the influential role French and foreign public opinion could play'.[27]

Separatist insurgents seem almost as unlikely as revolutionaries to be successful through any sort of pressure strategy, let alone one that includes terrorist-mode attacks on their local state's equivalent of a homeland territory. This homeland or heartland territory could be readily identified in geographical terms in the case of the IRA insurgencies against the British state; it was simply the British or English 'mainland' lying across the sea from Ireland. In that sense little was changed by the 1921 Treaty which ended the IRA's insurgent campaign for an independent Ireland. The Treaty's compromise political settlement produced a partition into a) the Irish Free State that incorporated the Catholic south of Ireland and b) the Protestant-majority Northern Ireland that remained a part of Britain – and that later versions of the IRA would fight to separate from Britain and unite with the rest of Ireland. But these IRAs would differ from the first in the way they carried the fight to the mainland, notably the readiness to attack civilian targets, and this seems to be one of the reasons for their relative lack of success when compared to the first IRA's pursuit of a pressure strategy. For the strategic planner of the 1919–21 insurgency, Collins, showed a much greater understanding than later IRAs' commanders of the operational requirements and political calculus of a pressure strategy, especially the importance of public opinion:

> The Collins IRA did nothing to alienate but everything to enhance that popular support. There was thus no indiscriminate bombing of civilians, indeed a plan to machine-gun the assembled ranks of MPs in the House of Commons, from a vantage point in the visitors' gallery, was countermanded by Collins. In this way the Irish struggle for independence between 1919 and 1921 came to be supported as legitimate, *even in England*, by Press, politicians and people alike.[28]

In 1939 the British mainland was subjected to a bombing campaign by a new and less politically realistic version of the IRA, which rejected the creation of a Northern Ireland united to Britain rather than to the Irish Free State. However, these Northern-Ireland separatist insurgents seem to have been quite discriminatory in their numerous bomb attacks, which were aimed at economic and symbolic targets rather than civilians and therefore caused only seven deaths.[29] One interpretation of the bombings is that presumably they 'were designed to cause sufficient panic and fear within British society to force the government to open negotiations on the ending of partition' between Northern Ireland and the Irish Free State, but another interpretation is that the bombings were designed to produce pressure for negotiations by inflicting economic and emotional hurt rather than by causing a Douhet-like panic and fear.[30] Whatever the intent, the actual effect of the bombings was to produce a hostile reaction from the British public and a new piece of legislation, the Prevention of Violence Act, that allowed the police to deport IRA suspects.[31] Thanks to such repression by the British government *and* the Irish government, the campaign petered out in the following year, when the British were also subjected to German aerial bombing that made the IRA time-bombs seem like pin pricks in comparison and relegated them to historical obscurity.

In the mid-1970s yet another version of the IRA, the Provisionals, engaged in another Northern-Ireland separatist bombing campaign on the British mainland – carrying out more than 130 bombings in 1973–4.[32] In true pressure-strategy fashion the Provisional IRA intended 'to affect British opinion as a lever on the politicians', believing that a) British opinion would be affected by 'a disproportionate fear relative to the actual damage caused' by the bombing and b) 'the feelings of insecurity would translate into political pressure for withdrawal' from Northern Ireland.[33] However, the PIRA realised that civilian casualties would likely produce outrage and hostility rather than fear and insecurity, so a telephone-warning system was used to reduce the casualties from time-bombs planted near civilian targets. While the 1939 campaign may have been aiming to inflict hurt rather than terror, the new campaign was clearly aiming to cause something closer to a Douhet-like panic but without the civilian casualties – a relatively bloodless version of terror bombing.

But the perhaps inevitable local-level bungling or 'inaccuracy' would produce results that were little different from a bloody form of indiscriminate bombing, as was epitomised by the 1974 'Birmingham pub bombing' that killed 21 people:

> In Birmingham a warning was given but far too late to have any effect, for the bombs went off just seven minutes later. ... The Provisionals lost a great deal of support because of these bombs, both immediately and in the long term. For years afterwards, the reason given by many on the Left for no longer supporting the IRA was, quite simply, 'Birmingham'.[34]

Not surprisingly, this debacle produced a 'popular perception' of the PIRA as being 'psychopathic' and resulted in new legislation that increased the police's powers: the Prevention of Terrorism Act.[35]

Nearly two decades later the PIRA carried out another and much more politically skilful bombing campaign against civilian targets on the mainland. It had been pursuing a political-attrition pressure strategy that avoided arousing fear or insecurity among British civilians but the early 1990s saw a 'renewed emphasis on attacks in England' that now sought 'to overload the British economy by disrupting daily life through the bombing of shopping thoroughfares, central London stations and so on'.[36] Although the new campaign could not exclude debacles arising from bungling inaccuracy, it was clearly seeking to inflict economic hurt rather than create even a bloodless version of Douhet-like panic. Thus its most successful attack was the 1993 bombing of the City financial district of London, which resulted in financially very costly destruction and disruption but – thanks to the bomb-warning system – caused only one fatality.[37] It was the sort of bomb damage that the Zeppelin airships would have hoped for in 1915, when the City financial district had been their key target, and this attack seems to have produced some significant pressure on the British government.

However, the political context had changed, too, with the PIRA displaying a new-found political skill and realism. This would be particularly evident in its 1994 declaration of 'the cessation of military operations' and in negotiating the 1998 Belfast Agreement political settlement that gave these separatists some political respectability and symbolic successes.

Similarly, the Tamil-separatist LTTE 'Tamil Tigers' carried out some attacks in the Sri Lankan state's heartland territory as part of its pressure variant of the maoist-model preparation strategy that was described in Chapter 3. These bombings were indiscriminately imprecise attacks on political leaders and economic targets. Hopgood notes that such imprecise assassinations and attacks on 'infrastructure' can result in heavy civilian casualties (such as the more than 20 killed by the assassination attempt against President Kumaratunga or the more than 80 killed by the attack on the country's Central Bank) even though these civilian casualties were unintended and a form of collateral damage.[38]

However, the Tamil Tigers' attacks usually attracted international attention not because of their civilian targeting but because of their suicide-attack delivery system, which is discussed in Chapter 7. The Tigers were described by Pape as the 'world's leader in suicide terrorism', accounting for 75 of the 186 cases that he had identified in the period 1980–2001, but his list of their attacks included many combat operations and few, if any, attacks that were deliberately aimed at killing ordinary civilians in the state's heartland.[39] Similarly, when Pape pointed to the Tigers as a clear-cut case of suicide terrorists winning concessions from the enemy state, namely the 1994–5 negotiations, he also acknowledged that these terrorists' suicide bombings were only a part of a wider campaign which was inflicting 'punishment pressures' on the Sri Lankan state through 'conventional' as well as suicide attacks.[40]

Clearly the political pressure on the Sri Lankan government to negotiate a settlement was largely due to the successful maoist-model insurgency in the Tamil region of the country – not the suicide attacks on civilian targets in the state's heartland. And in fact the LTTE Tamil Tiger case was used by Abrahms in 2006 as evidence of 'why terrorism does not work', because his focus on targeting rather than delivery-system led him to categorise the Tamil Tigers as being more guerrilla than terrorist. His research into the effectiveness of terrorism included all 28 of the groups that had recently been designated 'foreign terrorist organisations' by the US State Department in accordance with its targeting definition of terrorism, namely that they had attacked non-combatant targets.[41] However, he discovered that on only three occasions had any of these groups been wholly or partially successful in achieving one of its policy objectives. What is more, all three successes had been achieved by a subcategory that he classified as 'guerrilla' terrorist groups, because they attacked mainly military targets, and *none* of the successes had been achieved by what he termed 'civilian-centric' terrorist groups, which 'primarily attack innocent bystanders and businesses'.[42] As would be expected, he had categorised the Tamil Tigers as having achieved a partial success in their policy objective of establishing a Tamil state and as being mainly military in their targeting and therefore belonging to the guerrilla sub-category.

The failure of the civilian-centric, civilian-targeting groups to achieve their policy objectives is further circumstantial evidence that insurgent bands lack the capability to instigate a Douhet-like panic or to inflict a decisive amount of hurt upon a civilian population. On the other hand, Abrahms also noted that terrorist groups are often able to achieve 'intermediate' objectives, particularly the attention-getting and agenda-setting objectives that are described in the next section as being part of insurgents' propaganda strategy.[43]

Propaganda and provocation – typically terrorist strategies

Although the preparation and pressure strategies are the most likely way that insurgents will achieve their political goals, the typically terrorist propaganda and provocation strategies have often been used by insurgents – if only in pursuit of supplementary or intermediate objectives. The propaganda and provocation strategies are described as 'typically terrorist' because they are pursued typically a) through the terrorist, civilian-targeting mode of insurgency and b) by insurgents who are categorised and denounced as 'terrorists'. These two features can in turn largely be explained by two factors. First, the targeting of civilians may well appear to increase the likelihood of these strategies being successful, even if the killing of civilians results in outrage and hostility. Secondly, propaganda and provocation are strategies that even weak insurgencies can carry out with some hope of success and, as is often emphasised, 'terrorism is the weapon of the weak, who lack numbers or conventional military power'.[44] Thus weak bands tend to adopt these two strategies and also tend to resort to the terrorist mode of operations. But of course this combination of tendencies is partly due to bands assuming that terrorist targeting increases their propaganda or provocation strategy's likelihood of success.

Insurgents' use of propaganda and provocation is just as complex and multifaceted as their pressure or preparation strategies. Part of the complexity arises of course from their inherently supplementary or intermediate nature; insurgents cannot hope to achieve their revolutionary, separatist or expeller political goals solely through propaganda and/or provocation. Sometimes they have been supplementing one of the two main strategies, preparation or pressure, but more often they have been aimed at an intermediate objective that is intended to precede and assist the insurgents' pursuit of one of the main insurgent strategies – or indeed some 'other means' than stateless warfare of achieving the insurgents' political goals. These possible 'other means' have included organised non-violent action, revolutionary insurrection, democratic elections or international intervention. For example, if the provocation strategy succeeds in provoking a politically counterproductive response from a government, this may lead to national or regional elections being won by a party that will further the insurgents' political goals.

The classic historical case of electoral success being based on a counterproductive response is the 1916 Easter Rising by Irish separatists. Whether by design or accident, this unsuccessful armed uprising provoked the British authorities into politically counterproductive repression. 'In all, military courts tried 160 of the rebels, and 15, including the signatories of the Declaration of Independence, were eventually executed'.[45] As might have been expected, the creation of martyrs shifted the political balance from the Home Rule moderate separatists to the Republican secessionists. 'After the executions, Republican organizations such as Sinn Fein increased dramatically in strength, and real power moved from Redmond and the Home Rulers to the Republicans'.[46] In the 1918 general elections to the British parliament, the explicitly secessionist Sinn Fein won 73 out of the 105 Irish seats or constituencies and so had a clear electoral mandate from the Catholic majority of the population. As Sinn Fein and the British government would not compromise, a separatist insurgency would be fought in 1919–21 before a negotiated settlement was finally achieved. But the Sinn Fein electoral triumph ensured that the British government was receptive to a compromise settlement, which eventually involved a) the Protestant minority being protected by partition, with the resulting Protestant-dominated region of Northern Ireland remaining part of Britain, and b) the independent Irish Free State having to accept temporarily the colonial-like status of being a self-governing Dominion loyal to the British monarchy. In fact the IRA's three-year insurgency was a relatively cheap price for the separatists to pay considering how long insurgent separatists in other countries have had to fight – and usually without having achieved any lasting victory. What is more, many other separatist insurgencies have produced a much higher cost in human lives than the Irish total of about 2,000, which included some 650 members of the IRA and the 300 or more civilians who 'had been murdered or simply vanished'.[47] It seems safe to say that the toll would have been much higher and the struggle much longer if not for the electoral victory of 1918 that in turn was due to the intermediate success of a provocation strategy.

The propaganda and provocation strategies are not only complex but also multifaceted. The different facets of the propaganda strategy were highlighted by Crenshaw in her analysis of the political logic for using terrorism and of the strategic choice of methods and techniques. Her analysis of terrorism's propaganda value emphasised its attention-getting or agenda-setting function: governments 'can reject but not ignore an opposition's demands' if it has 'skilfully articulated' why violence is being used.[48] But she also pointed out that propaganda deeds may be intended to weaken the state by undermining the government's authority and demoralising its administrators, police and military.[49] These various facets of the propaganda strategy were evident as early as the 19th century, when populist revolutionaries assassinated Tsar Alexander II and various Russian administrative and police officials for propaganda purposes. A leader of the populists later argued that the killing of a Tsar or a Minister was the way in which the movement became something that counted with the peasantry, who were influenced by events rather than words.[50] But clearly there was a state-weakening facet to this propaganda by assassinations, and it was also evident in a much more recent example – the 2003–4 attacks by the Al Qaeda band in Saudi Arabia (QAP). Its short-lived campaign of attacks on foreigners and other high-profile targets had a dramatic effect on the Saudi government's prestige: 'tremendous damage was done to the reputation of the government and its national credibility and international standing'.[51] Even particular attacks or incidents can have a politically damaging effect. In the Saudi case, the escape of some members of QAP from police raids had the effect of 'discrediting the regime's claims to efficiency and providing security, an important factor in upholding its legitimacy'.[52]

It is true that less extreme forms of violent action can also make effective propaganda. The Tupamaros showed long ago in Uruguay 'that kidnapping well-known politicians and businessmen and keeping them in hiding for several months can be a more effective way of humiliating a government than shooting down a minister in the street'.[53] As is described in Chapter 5, these urban revolutionaries and their counterparts in Argentina also showed that Robin Hood activities can be a relatively non-violent but effective way of attracting attention, even if only by distributing stolen turkeys to the poor. Similarly, other Latin American insurgents of that era used relatively little violence when they pioneered the use of airplane hijacking as a propaganda tool; in 1963 a Venezuelan domestic flight was hijacked by FALN supporters, who dropped leaflets over the capital city before having the plane flown to Trinidad and asking for political asylum.[54]

Yet there has always been a tendency to accentuate the violent aspects of what has long been known as 'propaganda by deed'. In the 1870s two Italian anarchists wrote that the affirmation of principles by deeds was the most effective form of propaganda, and Brousse went on to coin the term 'propaganda by deed' in an article in the same journal – arguing that a practical demonstration had the advantage of presenting issues so dramatically that they could not be hidden, distorted or overlooked.[55] By implication such a deed would be violently dramatic and aimed at the state, which had long been the target of anarchists' hostility, and indeed anarchists would take centre stage in an era of assassinations of political and administrative leaders that lasted until the early 20th century. The era is best remembered for the assassination of a French President and a US President, but there were also indiscriminate bombings of civilians by anarchists acting as individuals, who invented the justification that 'there are no innocents' in the enemy's social group or society.[56]

In contrast, the era's other major source of assassinations, the populist Russian revolutionaries, remained very discriminate in their use of violence. As noted earlier, they confined their targeting to Tsar Alexander II and various Tsarist administrative and police officials. But more importantly, the bomb-throwing assassins supplied by the People's Will band and the Revolutionary Socialist party would not carry out attacks that might also kill the targets' children or wife or some 'strangers' travelling in the same railway carriage.[57] It seems that every historical era sees marked variations in the concern to avoid casualties among innocent civilians.

When a new era of 'propaganda by deed' arrived in the 1960s–70s, it was focused on agenda-setting violent behaviour but with a new interest in kidnapping or hostage-taking and in the technologically new venture of high-jacking airliners. Furthermore, the dramatic propaganda deeds of such bands as the Palestinian 'Black September' were clearly internationally oriented and this led to the era's most famous theory of terrorist activities – Jenkins's argument that 'terrorism is theatre' – to highlight their international propaganda orientation and reliance on the international mass media to transmit the drama to a far-off target audience.[58] These propaganda deeds apparently succeeded in winning attention and setting the agenda in such far-off places as the US. For Dershowitz has described how 'the dramatic nature of the acts got our attention' and led many US citizens 'to seek to gain a better understanding of terrorists' grievances and to address the root causes of the terrorism'.[59]

However, the internationalisation also meant that the audience became more diverse and required a more carefully tailored message. A Western historian of jihad has pointed out that modern jihadist insurgents' attacks have a propaganda aspect that has to be explicable to both the Muslim and Western audiences: 'so that the appropriate messages are brought home to both'.[60] And the attacks also have to be not so terrorist in their targeting that they produce a backlash and constituency costs similar to those arising from the careless use of a provocation strategy.

The strategy of provoking politically counterproductive repression, like the propaganda strategy, is inherently supplementary or intermediate and is typically carried out in terrorist mode. What Schelling described as the use of 'sheer terroristic violence to induce an irrational response' is the typical way of provoking a politically counterproductive response, and the deliberate targeting of innocent civilians is the epitome of 'sheer terroristic violence'.[61] For example, the Algerian FLN insurgents seem to have deliberately targeted women and children with one of the time-bombs they used in 1956 against the European settler population in the capital city Algiers. The bomb was planted in a milk bar that was a popular spot for settler families and 'was filled with children and their mothers' when the time-bomb was planted and when it later exploded without any telephone warning having been given to evacuate the building.[62]

Such terrorist-mode attacks successfully provoked the French authorities into a politically counterproductive response. They handed over anti-terrorist policing in Algiers to the military and allowed them to use torture to gain the information they needed to destroy the FLN's networks in the city. As Horne points out in his classic *A Savage War of Peace*, this widespread and systematic use of torture not only alienated moderate Algerian opinion but also 'in France the stunning, cumulative impact it had was materially to help persuade public opinion years later that France had to wash her hands of the *sale guerre* [dirty war]'.[63] Thus the FLN's attacks on civilians proved to be a very successful use of the provocation strategy, eliciting a politically disastrous response from the French authorities.

The Algerian case is also a good example of how a typically terrorist strategy can be combined with other strategies. As was described in Chapter 3, the FLN was officially pursuing a maoist-model preparation strategy but would eventually defeat the French state by an inadvertent pressure strategy. This was in fact an inadvertent pressure-*based* strategy that used a supplementary provocation strategy to win an important campaign in this long and bitter war. Although the provocation indirectly led to an increase in the political pressure on the French government, this pressure was the result of a disastrous response that the authorities could have readily avoided and should be viewed as a victory won by a supplementary campaign strategy of provocation rather than as merely an aspect of the pressure strategy that won the war for this expeller insurgency.

Separatist insurgents, too, have a long history of seeking to win political victories against the enemy state by provoking it into a politically counterproductive response. As long ago as the 1880s a Fenian, Irish-nationalist bombing campaign in London was apparently aimed at provoking countermeasures from the British government that would arouse the fighting spirit of the Irish people as well as winning international sympathy for the Irish cause.[64] And when Taber argued that terrorist attacks could provoke a 'counter-terrorism' that served the attackers' cause 'better than any stratagem the rebels themselves could devise', he pointed out that the counter-terrorism of the Black and Tan units deployed in Ireland in 1920–1 had shocked even the British public.[65] In fact the Irish example may have been an early, separatist rather than expeller case of what Merom has described as luring an enemy democracy into brutal behaviour that will increase its liberal constituency's moral opposition to the war.[66]

But the Irish example also highlights how the provocation strategy can fail when employed by revolutionaries rather than separatists. When a faction of the IRA rejected the compromise 1921 Treaty settlement with Britain, they discovered that their fellow-nationalists who had been elected to govern the newly independent Irish Free State were not as vulnerable to provocation as a British government had been only a year earlier. The rejectionists' revolutionary insurgency against the new state's government learnt a hard lesson about the circumstances in which provocation will not be an effective method.

> They were convinced that if they could provoke the government into repression they could bring about a transformation of public opinion similar to that which had followed the Easter Rising and accompanied

> the Anglo-Irish War. They succeeded in the first of these aims. Like the British before them – and like most armies faced with the provocations of guerrillas – the government forces often hit out blindly and brutally … . But ultimately a large majority of the people were prepared to acquiesce in atrocities carried out by an *Irish* government, of the sort which they would not have accepted from the British.[67]

Some Latin American revolutionary insurgents, too, learnt the hard way that provocation of their own state was likely to be counter-productive. Perhaps the most striking case occurred in Brazil at the end of the 1960s. The leader of a Marxist revolutionary band, Marighella, had produced a soon-to-be famous *Minimanual of the Urban Guerrilla* that set forth a provocation-based strategy. The manual's section on Popular Support described how the urban guerrillas' rebellion would force the military government to intensify repression and use such measures as house searches, closing off streets, arresting innocent people as well as suspects and having the military carry out routine police duties. This transformation of a political situation into a military situation would 'make life in the city unbearable' and lead to political pressure on the military rulers to find a political solution by holding democratising elections – an attempt to 'fool the masses' that the guerrillas would have to counter.[68] But although the military regime met Marighella's expectations by intensifying repression, there would be no move towards democratisation until more than a decade after this repression had killed Marighella and virtually wiped out the country's many revolutionary insurgent bands.[69] Marighella's objective of militarising the situation had proved literally a 'dead end' for him and many other insurgents as well as being a political disaster for Brazil.

Furthermore, in the mid-1960s the government of neighbouring Venezuela had shown that the provocation strategy could readily be defeated without resorting to the militarised repression that Marighella sought to provoke. In 1962–3 revolutionary insurgents had planned to force President Betancourt's civilian government into using increasingly repressive measures that would cause him political problems. But this provocation strategy was defeated by a President who 'took great pains to show his respect for due legal process' and delayed the introduction of emergency measures until public opinion was in a receptive mood, as when an insurgent attack on a civilian excursion train aroused public opinion and was followed by emergency measures that led to mass arrests of communists.[70] Betancourt also successfully protected the 1964 elections against the insurgents and saw his chosen successor elected President of Venezuela. These elections were acknowledged by the revolutionaries and their sympathisers to have been a major political defeat for the insurgency – from which it never recovered.[71]

The Venezuelan and Brazilian cases represent the two main counterstrategies that were adopted to nullify revolutionaries' provocation-based strategy in the 1960s–70s. During this period urban revolutionary bands in not only Latin America but also Western Europe – notably the Red Brigades in Italy – sought to provoke their states into inflicting an unpopular amount of repression that would in turn provoke a mass insurrection. But a provocation-based strategy usually proved to be a 'self-destructive boomerang' that brought a repressive end to the insurgents.[72] In Western Europe a Venezuelan-like liberal-democratic approach was used but in Latin America a Brazilian-like militarisation was used, whether through the intensification, introduction, or re-introduction of military dictatorship. Only in the special conditions associated with the Somoza dictatorship in Nicaragua did the provocation of repression in turn stimulate support for the revolutionaries, including support from the middle classes, which would prove crucial in the 1979 military offensive.[73] Furthermore, this seems to have been an inadvertent use of the provocation strategy, for the Somoza dictatorship's response to any instances of rebellion, including the spontaneous small-scale insurrections that broke out in 1978, had been a more brutal and counter-productive response than the insurgents could have reasonably expected.[74]

A similarly inadvertent provocation strategy seems to have benefited the rural insurgencies in Guatemala and El Salvador in the 1980s. This was despite the fact that the regimes in both countries 'explicitly (if unsuccessfully) attempted to emulate the Venezuelan counterinsurgency "model"'; both regimes introduced elections and the Salvadoran President even compared himself to Betancourt.[75] But such an enlightened approach was undermined by the military in both countries pursuing their own versions of the Brazilian counterinsurgency model, with 'what can only be described as the grossest abuses of human rights by the armed forces, including frequent indiscriminate attacks on merely presumed regime opponents'.[76] Nonetheless, this politically counterproductive repression did not produce a Nicaraguan-like level of support for the revolutionaries and they would have to be content with the compromise political settlements negotiated in the 1990s.

Provocation is more likely to be an effective strategy in separatist rather than revolutionary insurgencies, especially when provocation is used to increase an ethnic or sectarian minority's sense of grievance or sense of identity. It is a long-recognised paradox that some ways of defending the state and society from insurgent attacks may be counterproductive, as 'the cost may be the alienation of a previously largely neutral population'.[77] And Byman has argued that *any* potentially successful response by the state against ethnic terrorism will have this counterproductive side-effect, because the response 'almost invariably requires an infringement of civil liberties' that will 'increase resentment of the government' and 'highlight divisions between suspect and non-suspect communities'.[78] In other words, it will highlight the separate identity of the ethnic/sectarian minority and increase or even create a sense of grievance against the government.

Of course the provocation is more likely to be effective if it involves a terrorist-mode attack on civilians, as a government is more likely to respond in a drastic manner if it is responding to attacks that arouse public outrage and hostility. On the other hand, separatists have to be careful that their terrorist-mode attacks do not have counterproductive side-effects for the insurgency rather than the state. Gurr's analysis of terrorism in democracies argued that terrorists' attacks could produce a counterproductive 'backlash' in the section of society that is supporting the insurgents, particularly if the acts of violence are directed against people whom the society considers to be innocent victims.[79] What has been described as 'constituency costs' is therefore one reason why such bands as the Provisional IRA or the Basque ETA have refrained from using extreme forms of terrorist-mode attack, notably using suicide attackers or targeting ordinary civilians, that are disliked by its hard-core supporters and by the wider section of society that the band claims to be representing.[80] For example, if ETA were to carry out 'systematic indiscriminate bombings, its constituency support among important sectors of the nationalist community would wane'.[81] So if the terrorist mode of provocation strategy is not used carefully, it may well become a double-edged sword that successfully provokes the state into politically counterproductive repression but also ensures that the insurgency does not benefit from the ethnic or sectarian divisions highlighted by that repression.

The provocation strategy presumably will also be affected by backlash and constituency problems when used in expeller insurgencies. And this may have affected the choice of targets in the key recent example of an expeller provocation strategy, Al Qaeda's 9/11 attack on the US homeland, for Bin Laden later emphasised that the attacks were aimed at military and financial targets rather than at such a distinctively 'civilian' target as a children's school.[82] The targeting clearly was also influenced by the attack's supplementary role as a propaganda deed, which sought 'to mobilize the support of Muslims whom the authors of the attack wished to win over to their cause'.[83] But 9/11 'was above all a provocation – albeit a provocation of gigantic proportions. Its purpose was to provoke a similarly gigantic repression of the Afghan population', similar to that inflicted by the Soviet Union in the 1980s, and 'to build universal solidarity among Muslims in reaction to the victimization and suffering of their Afghan brothers'.[84]

Al Qaeda's provocation strategy seemed to suffer an early demise when the US destroyed the Taliban semistate so quickly in late 2001 but then the strategy was given something of a reprieve by the 2003 invasion of Iraq. Not only did Iraq provide a new venue for Western repression that might unite the Muslim world but also the Iraq venture diverted the West's attention and effort away from the Taliban revival in Afghanistan and thus the long-term revival of the strategy's intended venue for Western counterproductive repression. Yet even in the long term the strategy is unlikely to succeed in reviving the Muslim solidarity and consciousness of the 1980s–90s. As Kepel argues, the strategy should be compared to that of the Italian Red Brigades, German Baader-Meinhof RAF and other leftist revolutionaries of the 1960s–70s who responded to the declining popularity of revolutionary consciousness by carrying out provocative terrorist attacks in the 'vain hope that the revolutionary consciousness of the masses could be revived and mobilized through a cycle of provocation, repression and solidarity'.[85]

On the other hand, the Al Qaeda provocation strategy may well succeed in stimulating local-level insurgencies in the almost inadvertent fashion that has recently been identified by Kilcullen in his theory of the 'accidental guerrilla'. These 'accidental' guerrillas in 'Iraq, Afghanistan, Pakistan, and elsewhere' are a local backlash to intervention by outside forces – Western and/or the local state's forces – that are seeking to destroy the safe haven that Al Qaeda has established in that locality; Al Qaeda's 'ideology tends to lack intrinsic appeal for traditional societies, and so it draws the majority of its strength from this backlash' against outside intervention in a traditional society's locality.[86] To what extent this is an inadvertent use of the provocation strategy is as hard to determine as whether it will be more successful than the 9/11 strategy. But both these forms of provocation are merely supplementary strategies that have been used in only a few campaigns of Al Qaeda's 'long war' or 'protracted war' expeller pressure strategy.

Finally, it must be noted that the use of provocation, propaganda or pressure are not the only strategies that may resort to the terrorist mode of operation. It has sometimes appeared in maoist-model insurgencies' preparation strategies, even if typically in a relatively moderate and discriminatory form. As was described in Chapter 2, Laqueur acknowledged that terrorism had been used by rural guerrillas as well as urban terrorists. And in his study of guerrilla warfare he pointed to the significant role that terrorism played in the Vietnamese communist insurgency of the 1950s–60s, especially when compared to the lack of terrorism in Mao's communist insurgency in China. The Vietnamese communists engaged in the 'assassination of village leaders, local teachers and other "dangerous elements"' to such an extent that by 1961 they had killed 'about ten thousand village chiefs in a country with about sixteen thousand hamlets'.[87] But the assassinations were a supplementary aspect of a predominantly guerrilla campaign and were at the moderate end of the civilian-targeting spectrum – far removed from the deliberate targeting of ordinary civilians. An analysis of the communists' use of terrorism in South Vietnam emphasised that their policy was to be discriminating and selective in the use of assassination and not to terrorise the ordinary people.[88] The assassinations were directed primarily at the administrative structure of territories that were fully or largely under government control, and its purpose was to enable the communists to initiate their social-political programme in that territorial area or village hamlet.[89] Unpopular officials were punished as enemies of the people, often with a notice attached to the body to explain why the official had been executed, but popular officials were kidnapped or were 'executed without fanfare'.[90]

It is true that there are cases where a maoist-model insurgency has gone further than the Vietnamese communists and has actually targeted ordinary civilians as well as local administrators. The Algerian FLN's anti-colonial insurgency used violence quite extensively against the Algerian population as part of its unsuccessful but continuing attempt to implement a maoist-model preparation strategy. However, the terrorist-mode violence directed against fellow-Algerians was again a discriminating and selective use of violence, and its extensive use was due to the FLN's attempt to maintain some control over the huge areas of Algeria that were in French hands and would be for years to come:

> The nationalists, with little more than a skeleton organization and the shrewd use of terrorism, could maintain their ascendancy even in areas of relatively high French military concentration. Failure to obey rebel directives often brought mutilation or execution to uncooperative Algerians ... [and] most Algerians refused to espouse

> Francophile views openly or to join any French-sponsored program as long as the slightest vestige of the FLN-ALN remained in the region.[91]

Clearly the extent and nature of the terrorist mode of operation in a maoist-model insurgency is related to the situation that the insurgency is facing. But clearly, too, it is an inherently secondary mode of operation for the maoist model's preparation strategy, which has its sights set on military conflict with the state. The preparation strategy is focused on creating a regular army of trained soldiers, not a gangster-like body of trained assassins and enforcers.

5
Comparisons with Criminality

When insurgents use organised violence to pursue their political goals and seek to make this a successful enterprise, they are using not only a strategy and a mode of operations but also an operational format. And as the format is readily categorised as bandit-like or gangster-like, this aspect of 'how they use organised violence' leads to comparisons with criminality rather than with states' warfare. The similarities in operational format between insurgents and professional criminals, whether bandits or gangsters, have been recognised since ancient times and is still noted in our present century. For example, in the aftermath of 9/11 the Chairman of the Joint Chiefs of Staff noted that 'destroying al-Qaeda would be like destroying an organised crime syndicate'.[1] And a US army officer operating against insurgents in Iraq in 2003–4 discovered that it was 'a task more akin to breaking up a Mafia crime ring than breaking up a conventional enemy battalion'.[2] Both these comparisons were pointing to insurgents' structural similarities to their gangster equivalents, such as the similar tendency to 'live apart and work together' instead of following the traditional bandit and military approach of living together as a social unit and true 'band of brothers'. But there are similarities in tactical as well structural techniques between insurgents and either gangsters or bandits and there are even similarities in the techniques used to pursue a propaganda strategy.

The insurgents' bandit-like and gangster-like operational formats are described in this chapter through comparisons with bandits' and gangsters' techniques, whether structural, tactical or propaganda. Differences in their techniques often reflect the tendency of bandits to operate in a rural environment and gangsters in an urban environment, and indeed the insurgents' operational formats are often described as the 'rural bandit-like' and the 'urban gangster-like'. But rural insurgents have sometimes adopted gangster-like techniques and urban insurgents have sometimes adopted bandit-like techniques, as when roadside bombs are used on rural roads and the Robin Hood propaganda technique is used in cities.

The similarities between insurgents and professional criminals reach a different plane when they engage in the *same* activities, such as insurgents engaging in the standard criminal activity of drug trafficking in order to fund their insurgency. So the chapter must also discuss and explore the nature of hybrid bands, which combine insurgent and criminal activities but usually give priority to one or the other kind of activity. On the other hand, it is important to emphasise how insurgents differ fundamentally from bandits and gangsters, who are professional criminals engaged in 'business by other means' – not politics by other means. The leaders of bandit or gangster bands are an illegal form of businessman and are focused on making a buck rather than a bang; the leaders of insurgent bands are an illegal form of political/military leader and are focused on the age-old political question of 'who rules?'.

In fact from political science's comparative perspective the leaders of revolutionary and separatist (and some expeller) insurgent bands are seeking power by the illegal means of political banditry or gangsterism. And therefore they are comparable to a) military officers that have seized or extorted power through the use of their country's military forces and b) the leaders of political parties that have misappropriated the public offices to which they were elected.[3] This comparison with criminality will be explored in the final part of the chapter, where the motive-means-opportunity forensic calculus of usurpation is presented as an introduction and bridge to Part II's analysis of the calculus of insurgency.

Comparisons with bandits and gangsters

Insurgents' enemies have often not merely compared them to bandits but actually labelled them as bandits. Chiang Kai-shek's 1930s campaigns against the Chinese communists were actually titled Bandit Suppression campaigns; more recently a Turkish Prime Minister of the mid-1980s described the PKK as 'only a bunch of bandits'; and in 2000 the Russian President referred to the Chechen insurgents as 'terrorists and bandits'.[4] Even the insurgents have at least recognised the similarities between banditry and their own operational format. Che Guevara's famous handbook on guerrilla warfare acknowledged the similarity with banditry in tactics and in all other characteristics except the guerrilla's need of popular support.[5]

Academic experts have said much the same. Hobsbawm pointed out the operational similarities between bandits and guerrillas, and also noted that some bandits are in fact no less dependent than guerrillas upon the active and passive support of the local population.[6] Laqueur's study of guerrilla theory and practice went even further and suggested that the 'ecology' of guerrilla warfare and banditry 'is identical to all intents and purposes'.[7] More recently, Kalyvas indirectly highlighted the difference between insurgents' stateless warfare and warlords' semi-state warfare when he contrasted bandits with warlords: whereas bandits 'must hit and run in order to survive, warlords levy taxes, administer justice, maintain some degree of order, and generally assume the burdens of government in the areas they control'.[8]

The most enlightening comparison between bandits and rural insurgents is based upon Hobsbawm's classic analysis of social banditry. These bandits establish a base in the forests or mountains, flourish in remote or inaccessible areas and in a few cases, though only in Chinese history, have expanded into a whole region and led a successful revolt against the monarch.[9] However, 'social' bandits differ from the common variety of bandit in being rebels against oppression as well as poverty. They do not prey upon the peasants but instead upon the lords or other outsiders and in fact 'are considered by their people as heroes, as champions, avengers, fighters for justice' – Robin Hood is the 'international paradigm of social banditry'.[10]

Consequently, social banditry has a propaganda strategy and techniques that beg comparison with those used by insurgents – urban as well as rural. For example, what Hobsbawm described as the 'avenger' type of social bandit carries out actions that show how 'even the poor and weak can be terrible' and can exert some degree of power, because inflicting death and pain is 'the most primitive and personal assertion of power'.[11] This is an aspect of propaganda by deed that is not commonly discussed but has been displayed quite recently by videos posted on the Internet by Iraqi insurgent groups. Benjamin and Simon pointed out that 'these new images of Muslims destroying military targets and enjoying dominance over their captives must be inspiring' to an Arab and Muslim audience that is used to television images of their people being 'routinely humiliated'.[12]

Another aspect of propaganda by deed that is not commonly discussed is the propaganda that takes the form of altruistic or ethical actions, as is epitomised by the 'noble robber' type of social bandit. This type adopts a Robin Hood image and approach that shows not only how 'poor men need not be humble, helpless and weak' but also that it is possible for them to achieve not merely vengeance but *justice*, as is shown by the bandits robbing the rich and giving to the poor and by the bandits' moderation in the use of violence.[13] The Robin Hood image was adopted by some South American insurgents of the 1960s–70s, notably by urban revolutionary insurgents. For example, in the Tupamaros' early years their 'reputation as romantic, benign revolutionaries was based on their restraint and on their Robin Hood-style actions to benefit the poor', as when they stole a lorry-load of turkeys and chickens on Christmas Eve 1963 and distributed them to the poor in the city's slums.[14] In the early 1970s the urban insurgents of neighbouring Argentina, too, adopted a Robin Hood approach, such as stealing a van-load of chickens and donating them to a strike fund.[15] However, the Argentinean bands took Robin Hood redistribution to new heights:

> Robin Hood actions were important as linkage with the poor. Among those were giveaways of stolen milk, meat, blankets, and toys in slums and the donation of 154 ambulances purchased with money extorted from Ford Motor Company. The kidnapping of Stanley Sylvester, British consul and manager of Swift and Company in Rosario who had recently laid off 4,000 workers, was a major coup; in exchange for his release, the company rehired the workers with compensation, improved working conditions, and distributed food in the city's slums.[16]

Finally, there was a variant of the 'expropriator' type of social bandit who believed that daring robberies would have an inspirational propaganda effect. This technique of propaganda by deed assumes that 'since men are good, brave and pure by nature, the mere sight of devotion and courage, repeated often enough, must shame them out of their torpor'.[17] Although the heroic robbery has never been popular with insurgents, a rather similar propaganda technique – again focused on the inspirational effect of heroic example – has been used by insurgents since the 1980s. Suicide attackers have been depicted as inspirational 'martyrs for the cause' by the Tamil Tigers, Hezbollah, and Palestinian insurgents, who were therefore 'signalling' to their potential supporters as well as 'signalling resolve and strength to the enemy' – which is apparently a 'side-effect' of using the suicide attack.[18]

However, the most important similarity between social bandits and modern rural insurgents is their inability to find a physically inaccessible refuge. Hobsbawm noted that improvements in transport and communications were often sufficient to produce a marked decline in banditry, and indeed 'on the whole social banditry is a phenomenon of the past, though often of the very recent past. The modern world has killed it'.[19] Much the same thing was said about rural insurgency in the 1970s, and although these notices of its demise proved to be premature, in recent decades it has indeed been overshadowed by urban insurgency – the counterpart of gangsterism rather than banditry.

The urban gangster differs from the rural bandit in dispersing his band among the urban crowds. It comes together only for 'work activities' and then often only in small groups or even as individual operators with a particular task or assignment to perform. Likewise, the urban gangsters' organised crime differs markedly in method from the hit-and-run raids and ambushes of rural bandits. For the bank robberies and other 'hold ups' that are the urban equivalent of bandit raids and ambushes are not typical of gangsters' organised crime. Their most important and certainly most lucrative activities are illegal businesses, such as gambling, prostitution, loan-sharking and the drug trade.[20] And the violence that is associated with these businesses is that of gangland wars with competing organisations or the use of 'hit men' to carry out businesslike assassinations of competitors, debtors, untrustworthy employees or obstructions to better business. The violence is therefore carried out with cool-headed premeditation and against almost self-selected targets, who are typically not the subject of any personal hostility or enmity.

This gangster-like form of violence has long been apparent among urban insurgents, who were engaged in assassination as long ago as the *sicarii* 'dagger-users' of Roman-occupied Jerusalem. A much more recent parallel was recognised by an Irish historian describing IRA killing of army and police personnel in Northern Ireland:

> 'We've nothing against them personally. It's the uniform we are after. It's just business,' said one IRA apologist to me in a comment that might have come straight out of *The Godfather*. Yet that spokesman and the bulk of his movement would have been highly insulted at being compared to the Mafia.[21]

Rural insurgents, too, have adopted a gangster-like approach to killing particular individuals, such as the assassination of South Vietnamese officials that was mentioned in Chapter 4. The similarity was noted in the early 1970s when Gann pointed to how guerrillas in the countryside, notably in Vietnam, adopted the style of city gangsters when carrying out the planned assassinations of individually targeted administrators.[22] However, this style has been more often adopted by urban insurgents and not only in the killing of civilians; they have also assassinated soldiers and police in hit-man style, as when IRA snipers shot British soldiers on foot patrol in urban areas of Northern Ireland.

Urban insurgents' propensity for gangster-like assassination has included using gangsters' explosive means of killing their victims. Gangsters have long used explosive devices as a means of assassination, whether to increase the chances of success or to reduce the risks faced by the killers. Laqueur noted that it was Al Capone and other Prohibition-era Chicago gangsters who perfected the car bomb, and it was the Sicilian mafia who first brought the roadside bomb to world attention when in 1992 the three-car convoy of a famous anti-mafia magistrate was destroyed by the remote-control detonation of explosives hidden in a motorway drainage pipe.[23] As Laqueur pointed out, it was insurgents who transformed the car bomb into a 'super torpedo', especially when driven by a suicide attacker.[24] But it was gangsters who first used the analogy with submarines and their torpedo weapon to describe the gangster-like form of violence; the Prohibition-era gangland slang described a hit-man or enforcer as a 'torpedo'.

In addition to these similarities in tactical techniques, gangsters and urban insurgents seem to have some similar structural techniques. In particular, there is some similarity between a) how gangster bands cooperate with one another to make their businesslike criminal enterprises more profitable and b) how urban insurgents bands have cooperated to make their operations more effective and efficient. While gangsters have sometimes engaged in gangland wars and assassinations, they have more often been engaged in businesslike cooperation: the equivalent of cartels, horizontally or vertically integrated business alliances and the promotion of joint ventures. To take a famous example, Al Capone's distinctive business strategy was to enter into one-off joint ventures with specialists in liquor distribution, gambling and prostitution.[25]

Similar forms of business-style cooperation are evident among urban insurgent bands operating in the same country or even internationally. For example, the five major bands operating in Argentina in the early 1970s covered a wide range of the revolutionary ideological spectrum, as they included Trotskyist and Leninist variants of Marxism as well as variants of Peronism, but they cooperated in a businesslike manner that would have been envied by a syndicate of mafia bands. 'While each operated independently, with some territorial division among themselves, there was considerable exchange of information among the groups as well as some coordinated operations'.[26] As for joint-venture partnerships, the Al Qaeda band has outshone Al Capone in its cultivation and use of this businesslike cooperative relationship. Al Qaeda band leaders spent much of their time 'selecting which of the myriad requests for assistance they would grant. The requests – for money, expertise, advice and other logistical support – came from everywhere, from Morocco to Malaysia'.[27]

The virtually global nature of these various requests and relationships is symptomatic of another important similarity between modern urban insurgency and organised crime – they have become more international in their activities and personnel. In the 1990s the growth of transnational organised crime became such an issue that there were United Nations' conferences devoted to the problem, culminating in the drafting of a UN Convention Against Transnational Organised Crime.[28] Even an academic expert who has serious doubts about the degree of control and coordination achieved by transnational gangsters acknowledges the existence of such 'gangster groups' as the Mafiosi, the Cammorrista, the Triads, the Colombian drug cartels and the Mexican drug traffickers.[29] And there seems little doubt that in recent decades it has become more common for a band of gangsters or insurgents to operate transnationally and perhaps draw its membership from more than one or two countries.

This is true of the Al Qaeda band as well as of the whole jihadist expeller insurgency that this band may seem to embody, represent or lead. It is important to distinguish between 1) an international insurgent or gangster band, 2) an international *version* of insurgency or organised crime, and 3) an international *type* of insurgency or crime. For example, the mafia version of organised crime spread from Sicily to the US not through the trans-Atlantic expansion of particular mafia bands or 'families' but through immigrants from Sicily pioneering in a new 'land of opportunity' the mafia version of organised crime.[30] They were so successful that soon US newspapers applied the label 'mafia' not just to a particular version of organised crime but also to the whole Italian subtype of organised crime or to the whole gangster type of criminal activity – and in the 1950s–60s 'public opinion again began to confuse the mafia with organized crime per se'.[31] In the 1990s–2000s there has been a similar tendency to confuse the Al Qaeda band with the whole jihadist version of expeller insurgency, and even with Islamist revolutionaries or separatists, in a way that is actually analogous to confusing a particular mafia family with the whole mafia version of organised crime.

On the other hand, this tendency towards simplification is very understandable considering the complexity of the jihadist phenomenon. For example, in his study of members of jihadist groups Sageman mentioned a large and diverse range of Islamist bands in different parts of the world: in the central Arab region there was the Egyptian Islamic Jihad, the EIG/GIE and 'various smaller Yemeni organizations'; in Northwest Africa there was the Algerian GSPC, Moroccan Salafia Jihadia and 'various smaller violent jihad organizations'; and in Southeast Asia there was the Jemaah Islamiyah and Moro Islamic Liberation Front.[32] Some of these were Islamist revolutionary or separatist bands, not jihadist expeller bands, but there was sufficient ideological similarity with the expellers, particularly a shared antiwesternism and opposition to Western influence, for there to be some sort of alliance or working relationship with jihadist expeller bands. Of course some Islamist groups have prudently avoided being associated with the jihadist phenomenon, such as in the case of the ethnic separatist insurgency by Patani Malays in southern Thailand. Although this insurgency has had Islamist connotations, Kilcullen points out that it has 'avoided attacking Western targets, spreading large-scale violence more broadly within Thailand, or accepting assistance from transnational terrorists'.[33] But given the number and variety of other Islamist groups that have been tarred with the jihadist brush, it is hardly surprising that Western public opinion has simplified the complex jihadist phenomenon by viewing this collection of insurgent bands as an inter-linked extension of the most prominent band in the collection. And ever since Bin Laden's 1996 declaration of war against America and his 1998 proclamation of jihad by the World Islamic Front, the Al Qaeda band has certainly been the most prominent.

Yet even after its late 1990s rise to prominence, Al Qaeda did not seek to take over or control the other jihadists but instead continued to work within a decentralised, complex structure of what Sageman describes as 'multiple alliances among separate organizations in the jihad'.[34] The 9/11 Commission noted that Bin Laden 'selectively provided start-up funds for new groups or money for specific terrorist operations' but 'it is unlikely that al Qaeda was funding an overall jihad programme' before the 9/11 attack.[35] Operationally, too, the Al Qaeda leadership sheltering in Taliban-ruled Afghanistan preferred indirect participation in joint-venture operations and, apart from the attacks on US embassies in East Africa, the Al Qaeda band did not become directly involved during the late 1990s in even the planning of operations – preferring instead to provide its allies or joint-venture partners with some training, seed money and logistic support.[36] Such an assortment of joint ventures and multiple alliances added up to what has been described in the case of the mafia as the presence of organisation rather than *an* organisation.[37]

Another structural similarity between mafia gangsterism and many urban insurgencies is the tendency for any sizeable band to be held together by hierarchical, vertical links between the leadership and individual members rather than by lateral, horizontal links between the members of the group. This structure of 'vertical dyadic' relationships – accompanied by secrecy about membership of the group – was adopted by the 19th century Sicilian mafia and inherited by urban mafia bands in Sicily and the US.[38] A similar structure was adopted by urban insurgent bands as a way of reducing the damage that can be inflicted by informers, undercover agents or information extracted from captured members of the group. For example, the Provisional IRA decided in the later 1970s that its military-style structure of brigades, battalions and companies was 'an outmoded pattern which was proving susceptible to penetration' and should be replaced by a compartmentalised, cell-based organisational structure and by what was described as 'a return to secrecy'.[39] The new structure was therefore composed of specialised four-man cells whose only contact with other members of the PIRA was the vertical-dyadic link between cell leaders and brigade-level officers. Similar vertical-dyadic cell-based organisational structures were adopted by urban insurgents in more recent times and in other parts of the world. In the 1990s this structure was being used in South-east Asia by the Jemaah Islamiyah and being espoused by Al Qaeda in its Afghanistan sanctuary.[40] In the 2000s it was also very evident among the diverse insurgent bands operating in Iraq. Although tribal-based and other local bands were relying on clannish social norms for security, the larger and more hierarchical bands used a cell-based structure of functionally specialised cells and of procedures that insulated the organisation's leadership from the operational cells.[41]

In fact the leaders of the band can be insulated from having to give orders directly to their operatives if the cell-based vertical dyadic relationships include a multilayered hierarchy or chain of command. The heads of mafia families have long used a multilayered structure to protect themselves from law-enforcement; in one fictional mafia family 'nothing could be traced to the top' because there were three layers or links in the chain of command between the Godfather 'and the operating level of men who actually carried out the orders'.[42] Similarly, the PIRA leaders used a multilayered hierarchy to protect themselves from the British security forces; the latter 'could rarely catch the "Godfathers", who took great pains to distance themselves from explosive weapons or operations of any sort'.[43]

However, in recent times there seems to have been some shift towards a horizontal *network* structure of hubs and nodes rather than a vertical hierarchical structure of ranks or layers exercising authority over subordinates. The flatter and looser internal structure is analogous to the criminal network structures identified in the 1990s by Dutch criminologists studying the 'network mode of organisation' used in:

> present-day sub-legal activities such as producing and trafficking illicit products and delivering clandestine services. Such activities exist in a hostile environment and thus they need a capacity of rapid innovation, adaptation and avoidance in response to possible law enforcement interventions. According to simple Darwinist reasoning, in a continuously changing world the more flexible 'social life-forms' stand the best chance of survival.[44]

Indeed it has been argued that a network structure gives these criminal enterprises not only greater adaptability but also greater resilience to damage, as they can rapidly rebuild their operational capacity by an Internet-like reconfiguration of network connections and capacity.[45]

A similar structural tendency and accompanying adaptability and resilience seems to have emerged among the jihadists in the 1990s. In his analysis of the overall jihadist movement's organisational structure Sageman describes two of the most important clusters of jihadists as having taken the form of transnational networks. These two 'small world' networks were scattered throughout the Western and Islamic worlds but both networks were given some localised structure by 'lieutenants in the field', who were the hubs of sub-networks that included the 'operational field commanders' in charge of particular insurgent operations.[46] It is true that this network structure was *not* adopted by the many other clusters or organisations that formed the inter-linked and often inter-allied jihadist movement. For example, it was not adopted by the hierarchically organised Jemaah Islamiyah in Southeast Asia or even apparently by what Sageman terms the Central Staff of Al Qaeda in Afghanistan. But the two networks are particularly interesting as examples of modernisation and of an especially adaptable and resilient form of insurgent band or sub-band, such as the often-noted resilience that they show when their experienced leaders are captured or killed and have to be replaced by new leaders.[47]

Of course there is a limit to jihadist network structures' resilience to damage. Sageman pointed out that although they were not as vulnerable as hierarchical structures to leadership-decapitation, they are vulnerable to breaking down into isolated, non-communicating islands of nodes if too many of the network hubs are eliminated.[48] However, even if they are unable to rebuild themselves by an Internet-like reconfiguration of the network, they may be able to 'reproduce' themselves through the creation of new nodes, even if relatively isolated and localised nodes, that arise from a process akin to hydra-headed growth. Byman has described how a 'hydra-headed monster with no central control' was produced by the Indian government's crackdown on Sikh separatism in the 1980s, which led to a 'splintering into violent factions' that 'made it difficult for the government to stamp out violence'.[49] And the key feature of hydra-headed monsters is not the number of heads but the fact that whenever one of the heads is cut off by the hydra's antagonist, two *new* heads grow from the stump left by the severed head.

Something similar to a hydra-headed fragmentation seems to have occurred among the jihadists after the international crackdown on jihadist bands that followed the 9/11 attack. Sageman described the post-2001 jihadist attacks as being 'more or less sponsored' by what remained of Al Qaeda's Central Staff *or* being carried out by 'spontaneous' local clusters 'under its inspiration'.[50] And Burke pointed out that after 2001 there was a shift towards a new style of jihadist attack by new, spontaneously formed groups operating autonomously. Initially, there was a transitional period of old-style attacks led by jihadists, not necessarily members of Al Qaeda, who had returned from Afghanistan to various parts of the Middle East before or after the US counterattack of November 2001.[51] But from then onwards a new style emerged, with 'autonomous cells' being formed independently and 'conducting operations independent of any central hub or authority'.[52] Even the 2002 Bali attack by members of Jemaah Islamiyah was an example of this new style because while:

> they received some assistance and direction from more senior militants who had visited Afghanistan in the preceding years, the cell was largely composed of young men who had no previous involvement in terrorism ... [and] appear to have decided to go ahead with a campaign of violence directed at local Western targets despite the opposition of more moderate senior figures within Jemaa Islamiyya.[53]

Benjamin and Simon's work on the globalisation of jihad described the same phenomenon in similar terms. A 'new breed of self-starting terrorist cells' had emerged and were proving more difficult for governments to counter because the new groups were not only harder to penetrate but also had 'few of the links to other radicals that the investigators need in order to disrupt other cells and networks'.[54]

The most prominent examples of these 'self-starting' microbands were those that carried out the 2004 Madrid train attack and the 2005 London Underground attack. The former was carried out by a largely Moroccan-Spanish microband of about a dozen members who were self-recruited through friendship and kinship ties and apparently were linked to Al Qaeda only by listening to an Al Jazeera broadcast of a Bin Laden audiotape that included Spain in the list of countries which should be punished for participating in the war against Iraq.[55] In the case of the London attack, the four-member largely Pakistani-British microband had been self-recruited through friendship ties, but there appears to have been some contact with Al Qaeda during visits to Pakistan by the microband's two leading members – and a martyrdom video statement by its leader was issued by Al Qaeda.[56]

If the self-starting bands are considered to be part of Al Qaeda, this part seems analogous to the local-set structure of some modern US gangs. For example, the 'gangsters of Minneapolis tend to be affiliated with large national collectives such as the Bloods, Crips and Vice Lords' but this vertical connection is almost as loose as the horizontal links between local gang 'sets' within the city, who often have 'nothing to do with other sets that claim allegiance to the same gang'.[57] Analogously, a self-starting local set of insurgents could be affiliated with or merely claim allegiance to the Al Qaeda band as an international collective, with as weak a vertical connection as that created by a local set of gangsters when they adopt the national insignia of the Bloods.

Another important similarity lies in the microband/local-set's financial or employment structure. Some members of a local-set gang may not be professional or career criminals but instead are employed in a legal job or have another legal source of income. In other words, they may live 'for' their local set's gang activities but they do not live 'off' those activities in the sense of being paid or financially maintained by the local set or its leader. The same applies to members of a self-starter microband. In fact they are very likely to live 'for' but not 'off' their insurgency and indeed some may have to spend their own money on financing the microband's activities, as in the case of the Madrid-bombing and London-bombing microbands.

Yet both the insurgent and the criminal worlds are so complex and diverse that obviously many other examples of insurgent or gangster bands will have quite different structures from that of a local-set gang. In the case of organised crime, not only are there many networked modern gangs but also the traditional gangster structure is by no means extinct. In fact there seems to be a functional reason for the continuing survival of the relatively hierarchical structure used by the mafia and other traditional versions of gangsterism, such as the Triads and the Yakuza, for this structure is associated with gangs which tend to exercise a territorial monopoly and engage in 'organised extortion' that also performs a state-like policing function for the illegal activities in their territory.[58] Thus the Dutch criminal world of the 1990s seems to have lacked local-set gangs but certainly featured not only networked gangsters but also the more traditional types of gangster band. The 'more traditional father-son-like authority relations can still be found' and there was also some scepticism among Dutch law-enforcement authorities about whether the 'social network paradigm' used to analyse drug-trafficking holds true for 'the much tougher Turkish, Kurdish and Pakistani heroin traders'.[59]

Similarly, the jihadists included bands that can hardly be compared structurally to a local-set gang – in terms of their internal structure and/or of their vertical relationship to Al Qaeda. In terms of internal structure, the earlier mentioned Al Qaeda on the Arabian Peninsula (QAP) that became prominent in 2003–4 through a series of attacks in Saudi Arabia was much more organised than any local set. It was a 'hierarchical organisation' that included a formal leadership, council of advisers, various specialised committees and a military committee with an 'operations' section that controlled the cells or operational units.[60] The Jemaah Islamiyah was not only internally a similarly hierarchical and complex organisation but also vertically a well-established and independent band, not a newly formed local set of Al Qaeda. Similarly, Zarqawi apparently had established his own jihadist band in Germany, as a rival rather than local set of Al Qaeda, before travelling to Iraq in 2003 and becoming leader of the most prominent jihadist band operating in the country; it was not until December 2004 that he merged his band with Al Qaeda and was publicly recognised by Bin Laden as the leader of Al Qaeda in Iraq.[61]

The Iraqi expeller insurgency was in fact initially characterised by the bewildering structural as well as ideological diversity of its many insurgent bands. In addition to the jihadists who developed or joined bands in Iraq, there were three local influences on the evolution of this structural diversity: 1) 'some of the early insurgent groups or cells were made up of family members or neighbourhood friends who came together to fight the foreign presence', 2) thousands of personnel from the former regime's army, party, paramilitary and security services formed insurgent bands with quite a hierarchical structure and 3) some of these ex-regime personnel transformed other bands 'from a set of isolated units and cells into something more professional'.[62] Hashim pointed out in 2006 that while neighbourhood or tribal insurgents still tended to be relatively decentralised and horizontally structured, the larger and more geographically extensive bands tended to have a much more hierarchical structure:

> We once heard a great deal about the Iraqi insurgents belonging to decentralized networks. I believe that journalists and observers used the term because it sounded 'sexy' and because being 'decentralized and networked' seemed to be *de rigueur* for terrorist or insurgent groups. The reality is that you simply cannot be a wholly decentralized insurgent group and continue to exist for long or be able to carry out more than very limited operations in a limited geographical locale (e.g. one's neighbourhood or tribal lands).[63]

Consequently, out of this structural diversity an evolutionary trend towards organisation began to emerge. There was an 'increase in sustained cooperation and coordination' among various bands, which included joint operations as well as ad hoc collaboration for a particular operation.[64] This suggested that the insurgency was evolving, like the jihadist insurgency had during the 1990s, into a mafia-like structure characterised by the presence of organisation rather than *an* organisation.

But by 2007 the evolutionary path was moving towards consolidation of the insurgency into a particular organisation, as Al Qaeda in Iraq (AQI) was becoming the predominant expeller band. In part this was because of its competitive advantages compared to other bands in the expeller 'marketplace' but it was also due to the US forces' new counterinsurgency strategy of winning allies at the local level.[65] These were apparently intended to be allies against specifically AQI rather than against insurgents in general and so the range of local allies was extended to include not just tribal and neighbourhood leaders of groups but also tribal and neighbourhood insurgent bands and even bands comprising or led by ex-regime personnel.[66] With many insurgent bands becoming US allies, being destroyed by counterinsurgency campaigns or being voluntarily or forcibly absorbed by AQI, the tendency was towards the insurgency becoming a national-level campaign by a large sub-band of Al Qaeda – a larger version of the campaign that QAP launched against the state of Saudi Arabia a few years earlier.

So the focus returns to the international or transnational character of Al Qaeda and the comparison with similar features or trends in organised crime. However, no transnational criminal band seems to have elicited a comparable degree of loyalty from its national or regional sub-bands. Only military analogies seem appropriate when describing the willingness of AQI to sacrifice itself for the greater good of the organisation by tying down scarce infantry resources that might otherwise have been deployed against its headquarters in the Afghanistan region. And no gangster organisation could have conceived of such a globally oriented grand strategy, not even a hybrid gangster/insurgent band.

Hybrid bands

Hybrid bands are usually imagined to be insurgents engaged in gangsterism rather than gangsters engaged in an insurgency – as insurgent/gangster rather than gangster/insurgent hybrid bands. This is doubtless because so many insurgent bands have raised funds through the same activities as professional criminals use to 'make a living'. The insurgents have engaged in armed robbery, smuggling, kidnapping for ransom, the drug trade, the protection racket and various other rackets identified by Adams in his seminal work on the financing of terrorism.[67] Such activities would have been categorised by the Tupamaros as 'logistic operations', aimed at acquiring funds and arms, and these funding operations became a prominent part of other urban insurgents' activities in the 1960s–70s.[68] Marighella's *Minimanual of the Urban Guerrilla* included a subsection on 'bank assaults' and suggested that they should serve 'as a sort of preliminary examination for the urban guerrilla in his apprenticeship for the techniques of revolutionary warfare'.[69] He was aware that such activities might lead to the insurgents being mistaken for 'outlaws' but argued that their political identity could be preserved by a careful use of violence and not stealing from the poor.

But the *rural* FARC (Revolutionary Armed Forces of Colombia) provides the classic example of gangster-like 'revenue earning'. It began military operations in 1966 and has become not only one of the longest-living insurgencies but also one of the best known internationally – thanks to its kidnappings for ransom and its role in the Colombian drug trade. In fact its drug-related revenues have been so substantial (an estimated 60 million to 100 million US dollars a year in recent times) they have enabled FARC to operate more as a semistate than a guerrilla band since at least the 1980s, during which it doubled the size of its army to some 18,000.[70] On the other hand, its method of extracting revenue from the cocaine drug trade seems more akin to that of a state than a gangster band, except perhaps gangsters operating a protection racket.

Instead of actually taking part in the business, FARC policed and taxed the drug traffickers who began in the late 1970s to buy coca from peasant growers living in areas fully or largely under FARC control.[71]

The same approach is used by the other famous example of drug-funded insurgency – the Taliban in Afghanistan. Some five years after their semistate was destroyed, the Taliban were still able to levy a range of taxes on the drug trade. In areas fully or largely under their control they levied taxes of 10 per cent on the farmers who grow the opium poppies, up to 15 per cent on the laboratories that convert the opium into heroin, and 18–20 per cent on the drug traffickers in need of Taliban protection for their operations.[72] Its slice of the largest heroin trade in the world means that the Taliban earns even more than FARC from the drug trade, which also produces politically and administratively debilitating levels of corruption among the Afghan state's police and administrators.[73]

Nonetheless, FARC's exploitation of the drug trade was more fundamental because it led to the insurgency becoming an apparently hybrid entity. During the 1980s FARC not only increased its military strength but was also able to 'develop a stable economic infrastructure that could fund much of the organization's growth' – thanks to FARC having invested much of its drug-related revenues in 'licit enterprises in Colombia's major cities'.[74] In fact it was spending only about 10 per cent of its drug-related income on the military campaign, with the rest being invested in stock markets, companies and land.[75] Not surprisingly, FARC has been described as a 'hybrid organisation' that combines political and economic agendas.[76]

But Bovenkerk and Chakra's notion of hybrid 'organized criminal terrorists' is a potentially more useful conception of a hybrid insurgent band, even if there is little reliable evidence of terrorists having converted to organised crime:

> Various authors keep referring to more or less the same examples, the FARC in Colombia, Sendero Luminoso in Peru, guerrilla fighters in Chechnya, the Abu Sayyaf Group in the Philippines, the IMU in Uzbekistan and a few others. It is striking that none of them provides a thorough empirical analysis of any of these cases, and the evidence cited never goes much deeper than a good media account. This is not to belittle good journalism but even good reporters may miss asking the appropriate analytical questions.[77]

The right analytical questions to ask might be a) whether the revenue-producing criminal activities that had been grafted onto the insurgency are now taking priority over its political goals, b) whether they have actually become an end in themselves rather than a means to the military/political end, and c) whether the political goals have in fact been implicitly or explicitly abandoned and the band become gangsters rather than insurgents.

Hybrid bands have certainly arisen in the opposite direction, from insurgency being grafted onto criminality rather than vice versa. This means much more than seeking political goals that are a means to the end of protecting the gangster band and its business, as when the Italian mafia engaged in a series of car bombings of public buildings in an attempt to intimidate the state authorities into not passing anti-mafia legislation.[78] Hybrid bands seek broader or more altruistic political goals. Throughout history bandits and gangsters have sometimes taken part in separatist or expeller conflicts, such as in the recent case of Chechen gangsters providing recruits and resources for the separatist war against the Russian state.[79] Another recent example was the presence of drug dealers and other petty criminals among the membership of the microband that carried out the 2004 Madrid train bombings.[80]

Furthermore, the hybrids should be distinguished from cases of the neofeudal subtype of separatist insurgencies, which were described in Chapter 1 as being an insurgent equivalent of warlordism. Its new prominence in parts of Africa in the 1990s was reflected in the addition of a new, 'commercialist' type to O'Neill's typology of insurgencies. It already contained seven types, which had been defined by their political goals: the secessionist, preservationist, reformist, anarchist, egalitarian, traditionalist, and pluralist.[81] The commercialist type differed from these types because its 'main aim appears to be nothing more than the acquisition of material resources through seizure and control of political power' and 'such groups are essentially self-aggrandizing nihilists, although predators would also be an appropriate term'.[82] The African context was reflected 1) in his point that many examples 'have a narrow tribal or clan basis and are led by self-styled chiefs, warlords and the like' and 2) by his choice of example: the RUF of Sierra Leone.[83] Another possible example appeared in the same period but on the other side of the African continent – in Somalia's capital city Mogadishu. Although the factional leaders of clan-based bands had the long-term goal of gaining state power, they were associated with businessmen in joint-venture enterprises for the pillaging and on-selling of looted goods to such an extent that the competition for the spoils of looting joined with the competition for the spoils of office as a motive for inter-band conflict.[84]

This seems to be almost a hybrid version of the neofeudal insurgency and is almost as difficult to categorise as West Africa's 'roving' bands. These 'tend to resort to predation and pillaging while they are on the move' but apparently there is a difference between the bands that roam along the main roads, which are usually seeking to win control of the state regionally or nationally, and those that roam along bush paths and seek to produce an alternative political order within the territorial 'enclave' that they control – as virtually a roving semistate.[85] That a *roving* semistate seems a contradiction in terms is another indication of how the reality of insurgency seems much more complex than any conceptual analysis of this phenomenon.

The forensic calculus

The comparison with criminality is the appropriate place to introduce just such a conceptual analysis, namely the forensic-like motive/means/opportunity analysis and calculus of insurgency that will provide the theoretical framework of Part II. The insurgency calculus is an adaptation of an earlier application of the forensic approach to non-criminal (or politically criminal) activities by what was termed 'the calculus of usurpation'. It analysed, explained and predicted the modern means of usurping power, namely by a military seizing power through a coup or by a party either seizing power through revolution or misappropriating power through converting its election-won temporary hold on public offices/powers into a permanent tenure. The calculus of usurpation was in turn based upon a calculus of 'intervention' – by the military in politics – that had been developed in the early 1960s by Finer in his pioneering book on military coups and regimes.[86] His analysis and calculus not only used a combination of motive and opportunity but also recognised that coups were motivated by a mixture of motives, which could range from altruistic concern for the national interest to very prosaic concern for individual self-interest, and that there were inhibiting motives which prevented or deterred the military from intervening in politics.[87] Finer's calculus of intervention later had its scope and forensic analogies broadened by adding a 'means' variable to accompany the motive and opportunity variables and by applying the calculus to party as well as military dictatorships. Renamed 'the calculus of usurpation', it was then used 1) to predict the occurrence of military or one-party rule, 2) to explain why past and present examples have occurred and 3) to explain or predict the success or failure of *attempts* to usurp power in these ways.

When this forensic calculus is adapted and applied to insurgencies, it in turn seeks to explain or predict the existence – and success or failure – of insurgency in a particular country, region or global era. But as Goodwin has pointed out, categorising insurgencies as simply successes or failures does not adequately cover the range of outcomes:

> we need an additional category – besides 'success' and 'failure'– to characterize ... those that persist for many years or even decades without seizing power, but which maintain significant popular support. This last category includes those undefeated rebellions (like the Salvadoran and Guatemalan) that are terminated only through negotiated settlements.[88]

Adding a third category to cater for undefeated insurgencies is perhaps not as tidy as simply considering outcomes in terms of the *degree* of success – of how much success an insurgency achieves. This degree of success would range from immediate defeat at one extreme to rapid and cheap victory at the other extreme, specifically achieving the insurgency's political goals without a lengthy and costly use of organised violence. In between these two extremes lie a varied range of outcomes, including the two types of partial success mentioned by Goodwin: 1) the negotiated peace settlement that gives the insurgents an armistice amnesty and the opportunity of using peaceful, political means to achieve their political goals and 2) the persistence of an insurgency for many years, or even for decades, while still maintaining a significant level of support. Some persistent insurgencies have in fact become part of a country's 'routine politics' and a 'way of life' for the participants, as in the case of the decades-long ethnic separatist insurgencies in Burma or the FARC revolutionary insurgency that began in 1960s Colombia.[89] Whether this constitutes a greater degree of success than achieving a negotiated settlement is a moot point and would depend largely on the terms of the settlement, which have varied markedly between such cases as the Maoists in Nepal, the FMLN in El Salvador and the Provisional IRA in Britain.

The calculus makes explanations or predictions of this varying degree of success by making an assessment of the three variables, motive, means and opportunity, to determine whether each of them individually and the three of them collectively are of sufficient magnitude for an insurgency to be instigated and to achieve some degree of success. The collective or overall assessment is to some extent a *net* assessment, because the strength/weakness of an individual variable may counterbalance or compensate for that of another variable, as when a great opportunity compensates for insurgents' small means of exploiting that opportunity. Another 'net' aspect of the assessment is that the increased opportunity for insurgency may well arise through a decline or weakening of the insurgents' enemies, so that the assessment is actually being made in terms of the strength of insurgents versus the strength of their enemies. This sort of net assessment also appears in the other two versions of a forensic calculus. For example, Finer's calculus of intervention emphasised the government's lack of legitimacy as an opportunity for the military to intervene, and the calculus of usurpation includes the weakening of state power as a key opportunity for a party to seize power through a revolution.[90] Although such an assessment appears to complicate the calculus, it can actually simplify it by making a direct comparison of the insurgents' strength versus that of the state they are fighting against. And it also helps to highlight the effect of changes in the relative strength/weakness of the means and opportunity variables, especially how the changes may counterbalance or compensate for one another. In particular, the strengthening of the means of insurgency may well be counterbalanced by a strengthening of the state that has thereby weakened the insurgents' opportunity for success.

In addition to this lateral net assessment of one variable versus another variable, there is also a net assessment *within* each variable. For as the other two versions of the calculus pointed out, the motives for intervention or usurpation – or insurgency – are counterbalanced by inhibiting motives.[91] In the case of insurgency there are moral inhibitions against illegality and violence as well as the obvious prudent inhibitions arising from the fear of failure and of the repercussions of failure.

Finally, there are two different perspectives on the insurgency calculus – the analyst's and the participant's. The analyst's objective perspective belongs to the independent strategist or political scientist; the participant's subjective perspective belongs to the leadership and ordinary members of an insurgent band. However, there is some overlapping between the two perspectives, because the analyst's explanations and predictions implicitly assume that the insurgents engage in some sort of calculation of the variables, even if a merely instinctive assessment, before making such key decisions as whether to found or join an insurgent band and whether to continue with or abandon an insurgency.

This subjective perspective of the participants is similar to what Crenshaw described as an 'outlook' in her 1990 analysis of the logic of terrorism. She referred to a change in the terrorists' outlook – a 'combination of optimism and urgency' – that can arise 'when the regime suddenly appears vulnerable to challenge'.[92]And she analysed this regime vulnerability in a similar manner to assessing the means and opportunity variables. Both military and political opportunities were mentioned: a) the weakness of the regime's 'coercive resources', whether through overextension or through inadequate security precautions, and b) popular disapproval of the regime or a change in the 'climate of international opinion' that reduces its legitimacy.[93] As for the means variable, she pointed out that the regime's challengers gain an advantage from 'acquiring new resources', including such technological advances in weaponry as the 19th century invention of dynamite: 'science was thought to have given a decisive advantage to revolutionary forces' by providing them with 'such a powerful and easily concealed weapon'.[94]

Furthermore, Crenshaw noted that her approach was compatible with Waterman's explanation of collective political activity as 'determined by the calculations of resources and opportunities' – which seems very similar to the calculation of means and opportunity variables.[95] The similarity is evident, too, in the way he emphasised the importance of opportunity:

> Only if worthwhile opportunities for successful action present themselves will large groups, consisting of ordinary people with other responsibilities and values to protect, expend – possibly at some risk to themselves – time, effort and/or money in pursuit of satisfaction. ... All the misery in the world may not motivate a group to act against an authority it cannot defeat; even the most acute sense of injustice may not move large numbers of people to act together in a hopeless cause.[96]

Waterman also referred to equivalents of both the analyst's and the participant's perspectives on the insurgency calculus. He referred to 1) an analytical 'calculus' that seems to be an explanation involving 'significant changes in the resources and opportunities' and 2) the potential participants 'calculating their options' and engaging in a 'calculation' that includes resources, costs and the probability of success.[97]

So the calculus of insurgency is using an approach that not only has long been applied to studying non-democratic rule but also is clearly 'compatible', to use Crenshaw's term, with previous analyses of terrorist insurgency and collective political action. Although this compatibility seems to involve only the means and opportunity variables, the motive variable has been at least implicitly or tacitly included in these analyses.

For example, Waterman's description of potential participants 'calculating the probability of success' obviously implies a 'fear of failure' that may well be as powerful an inhibiting motive for potential insurgents as it is for potential participants in military intervention or some other form of usurpation. In all three versions of the forensic-like calculus – intervention, usurpation and insurgency – the fear of failure is an important inhibiting motive that reduces the motive to participate in that activity.[98]

In all three cases the analyst has to make a net assessment of the inhibiting versus activating motives that seems quite similar to the military type of net assessment, which has the 'problem of estimating the performance of one's armed forces against one's potential enemy'.[99] But net assessment of motive involves a more complex inhibiting force, which includes individual variations in fearfulness – the bravery/cowardice or risk-aversion factor – and variations in the type of failure that is feared. For there is fear of failure 1) in the tactical sense of the individual being killed, maimed or captured in insurgent operations or through the enemy state's countermeasures; 2) in the strategic sense of the whole insurgency being a failure and therefore wasting the insurgent's efforts, risks and sacrifice; and 3) in the opportunity-cost sense of the 'other responsibilities and values' that the insurgent might otherwise have protected, as Waterman noted when describing opportunity, or of the life that the insurgent might have otherwise enjoyed (and so an individual with the proverbial 'nothing to lose' is unlikely to be deterred by an opportunity-cost fear of failure).

In addition, the fear of failure is obviously related to the likelihood of failure. For the more likely it is that there will be tactical or strategic failure, the more the fear of failure will inhibit the motive to participate in these activities. Assessing the likelihood of failure will require a military type of net assessment – estimating the performance of the insurgents against the forces of the enemy state – but often will involve political as well as military factors, as when estimating a pressure strategy's likely degree of success. In general terms therefore the estimate is the equivalent of analysts' objective net assessments of means versus opportunity but it is being made from the subjective perspective or outlook of the potential participant and is focused on the likelihood of tactical or strategic failure.

The analyst's net assessment of motive, of inhibiting versus positive motives, will therefore have to incorporate some estimate of participants' subjective assessment of the insurgency's means and opportunity. But of course the analyst's objective assessment of the insurgency's means and opportunity may be very different from that of the participants, whether because of better information, more experience and training in making such assessments or simply the benefits of being able to take a more objective view of the process. From that perspective the analyst should be able to make a credible overall calculation of motive, means and opportunity that will offer a thorough and systematic explanation or prediction of an insurgency's degree of success. It is true that this series of calculations, net assessments and even estimates of subjective assessments is a complex challenge but it is no more complex than what has long been required by the calculus of intervention and has more recently been required by the calculus of usurpation. And as with those two forms of the forensic calculus, the bare bones of the calculus have to be fleshed out, as they will be in Part II (see Table 5.1), with detailed explanations of the variables and how they operate within the specific context of, in this case, insurgency rather than intervention or usurpation.

Table 5.1 The Insurgency Calculus

Motives +	Means: strengths/ weaknesses +	Opportunity: strengths/ weaknesses
1) positive motives	1) methods	1) physical – rural or urban
a) *political goals*	a) guerrilla/terrorist mode	2) political
b) unofficial, self-interest of band	b) bandit/gangster format	3) military
c) unofficial, self-interest of social group, e.g. tribe	c) strategy: i) preparation ii) pressure iii) propaganda iv) provocation	a) *enemy* military's method and capability
d) *individual* self-interest = selective incentives	2) capability	b) military-political control = 'fearful subjects' + 'local allies' in social support base
2) **minus** inhibiting motives	a) quality of leadership	c) passive support of 'silence' supplied by social support base
a) moral inhibitions	b) PIE i) Personnel: numbers/quality (including *motives*)	4) social ('political supporters')
b) *fear of failure* i) tactical ii) strategic iii) opportunity cost	ii) Information iii) Equipment (PIE built upon active support from social support base = recruits, information and logistical support)	

Furthermore, there are several major differences between the insurgency calculus and the calculus of usurpation. For example, the means variable is in both cases divided into capability and method, which in the case of the military is the coup method and in the case of a political party either the revolutionary or electoral-misappropriation method. But insurgency has a multifaceted method that includes not only the mode of operation but also the operational format and, most importantly, the particular strategy through which the insurgents plan to be successful (even if a different, inadvertent strategy actually brings them more success). All three aspects of method have already been explored in Part I as the various aspects of how insurgents use organised violence to achieve their political goals, so it is only the capability component of the means variable that will be explored in Part II's description and application of the insurgency calculus.

Part II will also be combining the insurgency calculus with an approach that has not been applied to usurpation – the environmental approach. It analyses the local environments in which insurgents have to operate and emphasises how these insurgency-related environments a) vary according to geographical place and historical time and b) have several different aspects or dimensions, not just a rural-urban physical dimension. The final chapters of Part II will therefore be analysing both the opportunity variable of the calculus of insurgency and also the local physical, political, military and social-political environments in which insurgents operate.

Part II

The Calculus of Insurgency: Motive, Means and Opportunity

6
The Motives of Insurgency

As was discussed at the end of Part I, the motive/means/opportunity calculus of insurgency seeks to explain or even predict the existence and degree of success of insurgency in a particular country, region or global era. Obviously the motive variable is the most important of the three in the calculus because without a motive no individual will ever start an insurgency or join an insurgent band, but with a sufficiently strong motive some individuals might start or join an insurgency even if there was no means or opportunity of achieving any degree of success. It is true that the description of the insurgency calculus in Chapter 5 depicted the motive variable as virtually dependent upon a net assessment of the means and opportunity variables, because if that assessment indicated that the insurgency was likely to fail, then the fear of failure would act as a powerful inhibiting motive that would override any motives to join the insurgency. But there are always likely to be some individuals with a sufficiently strong motive – or a sufficiently flawed subjective assessment of the insurgency's prospects – to start or join an insurgency that has no likelihood of achieving any degree of success. In that sense, motive is both a necessary and a sufficient condition for the existence of insurgency. It is also the crucial variable for explaining or predicting which individuals – and how many – become founders, leaders or ordinary members of an insurgent band.

Motive is also the most complex and difficult variable to assess. Several of the complexities or distinctive features of the motive variable were discussed at the end of Chapter 5. There are also several distinctive complexities or issues that arise from assessing the motive of insurgents, in addition to the problems involved in analysing any form of collective action – peaceful or violent, criminal or military. Finally, the chapter will examine the almost bewildering variety and mixture of motives that may be involved in the motivation of insurgents.

The problems of motive

The first issue is the relatively straightforward one of distinguishing between the motives of insurgents and those of their supporters. There is a crucial difference between members of a) the insurgent band and b) its social support base which supplies the band with active or passive support. (The various forms of passive and active support and the motives for supplying them are described in Chapters 8 and 9.) For the members of the band are those who actually carry out its organised violence, as distinct from those who supply the band with the various forms of assistance that it needs to continue operating as an effective force. However, this distinction may appear less significant than that between being a full-time or part-time member of the insurgent forces. For example, Petersen distinguishes between becoming a member of rebel guerrilla or regular forces and two lower levels or 'nodes' of rebellious role: a) directly supporting or participating in a locally-based armed organisation and b) merely attending a mass rally or writing graffiti opposed to the rulers.[1] The middle of his three levels of rebelliousness therefore includes not only supporters of the insurgency but also individuals who go beyond support and actually become a part-time local insurgent. But where the line is drawn between joining the band and supporting it is not as important as the fact that a stronger motive is needed if the individual is making a more dangerous and active contribution – a proportionality between motive and contribution that Petersen describes as 'triggering mechanisms' and 'thresholds' that move an individual to a particular level of rebellion or from one level to another.[2]

A second issue or complexity is the difference between motives a) for joining the band and b) for continuing to remain a member and carry out its organised violence. An example of the motive for continuing was provided by Sageman's rather provocative argument that friendship ties 'cultivated in the jihad' are so intense and significant that they may motivate jihadists 'to carry out horrific acts more easily than negative emotions do'.[3] A more conventional example is provided by a 1967 study of captured Viet Cong, which suggested that they had joined the insurgency for various other reasons than communist ideology but noted that these recruits had been indoctrinated *after* they had joined and that this may have reinforced their motivation to continue fighting.[4] In contrast, Petersen's analysis of why rebels continue with their roles included such a simple 'sustaining mechanism' as coercive 'threats', but he also pointed to 'irrational psychological mechanisms' that distort rebels' perception of the rebellion's prospects – and thereby reduce their fear of strategic failure.[5] Three of these psychological mechanisms are: a) an irrational assessment of future benefits because the individual cannot accept that past, 'sunk' costs were for nothing or were in vain; b) the subjective satisfaction produced by 'small victories', which prevents the individual from realistically assessing the overall course of the conflict; and c) 'wishful thinking' leading to the individual interpreting events as evidence of positive developments or prospects that, realistically, are very unlikely.[6]

A third issue is whether the fear factor in the fear of strategic/tactical failure is a psychological or a cultural factor. The key inhibiting motives that counterbalance the motivation to join or continue with the band are the 'fear of failure' inhibitions (see Chapter 5 and Table 6.1) and in turn their key variable is the bravery/cowardice or risk-aversion factor. If it is a psychological factor that varies as randomly as anger-management problems, then the assessment of motive will not be affected by the cultural background of the individuals being assessed.[7] However, it is the cultural rather than psychological factor that is emphasised by those who have some comparative knowledge, and especially hands-on experience, of insurgency or counterinsurgency. Thus Kilcullen's recent experience of countering both the Iraqi and Afghan expeller insurgencies led him to 'go on the record' about the differences in risk aversion between these two cultures:

> At the risk of reinforcing cultural stereotypes I would be remiss if I failed to record my observation that ... it was a rare Iraqi insurgent who loved the fight itself. The Afghan insurgents and former insurgents I have encountered *do* love the fight: they like to win and are certainly not averse to killing, but what they really like is the fight, *jang* (battle), for its own sake.[8]

At a more academic level, the cultural factor has been emphasised by Shultz and Dew in their account of the 'warrior' culture that continues to exist among tribes and clans in such conflict areas as Afghanistan, Iraq and Somalia.[9] They point out that as long ago as the 1970s Bozeman's study of war in non-Western traditional cultures was warning that some of them showed a more positive attitude to war than is found in Western societies.[10] If so, then presumably the average member of a tribe or clan with a relatively pro-war or pro-warrior culture would be less averse to risking life and limb in an insurgency than would the average member of a Western society.

A fourth issue or complexity is the difference in motivation between the band's ordinary members and its leadership, including the founding leaders. The ordinary members may well have a different motivation, such as acquiring some pay or loot, protecting the interests of their tribe or other social group, or achieving a less ideological political goal than the one being espoused by their leader. As Kalyvas has pointed out, the importance of ideological motivations in historical civil wars 'has been greatly overstated' because leaders' ideological pronouncements have mistakenly been viewed as indicative of the rank-and-file's motivations – there has not been sufficient acknowledgement of the fact that 'rebel motivations are diverse'.[11]

Fifthly and most importantly, there are the issues arising from the mixture of motives within the individual as well as within the band, especially the mixture of collective good and selective incentive. Of course the collective motive is associated with the insurgency's political goals, which are seeking to satisfy the political grievances of the insurgency's political supporters, as will be described in Chapter 9. But in addition to this political and publicised collective motive, there will be other collective and even individual motives influencing the members of the band. A similar mixture of diverse motives is also found in cases of usurpation of power, whether by militaries or parties, and the calculus of usurpation lists several different types of motive: 1) the national interest or an ideological interest and 2) various types of *self*-interested motive: a) the organisational self-interest of the military or party, b) the social self-interest of the military officers' or party members' social group, particularly their social class or ethnic group, and c) the individual self-interest of the members of the military or party.[12] In comparison, expeller and revolutionary insurgencies have national-interest or ideological political goals, while separatist insurgencies interpret their political goals' ethnic social self-interest as being the 'national interest' of their ethnic group. As for the various types of self-interested motives, they are the insurgents' unpublicised, unofficial collective and individual motives; they include the band's collective self-interest, the social self-interest of the member's class, ethnic/sectarian group or tribal/clan group and of course the member's individual self-interest, which might take such prosaic forms as greed for pay or loot (see Table 6.1 on Motivation).

The most obvious difference among these insurgent motives is the contrast between a) the publicised collective motive associated with the

Table 6.1 Motivation (Net Assessment)

Collective Motives	plus	**Individuals'** Selective Incentives	= Positive Motive minus	**Inhibiting** Motives
1) political goals/ grievances		1) **material** incentives a) 'greed' for material pay offs b) coercion through intimidation		1) moral inhibitions e.g. against illegality
2) unofficial, self-interest of the **band** itself		2) **social** incentives: rewards or sanctions 3) **affective** incentives		2) 'fear of failure' a) tactical (casualties) b) strategic (wasted sacrifice)
3) unofficial, self-interest of a social class, ethnic/sectarian, tribal or other **social group**		a) affection e.g. bunch of guys b) antipathy e.g. revenge 4) **moral** incentives a) coercive – guilt, losing self-esteem or personal identity b) positive – boosting self-esteem or personal identity 5) religious or secular immortality?		c) opportunity cost (something to lose)

insurgency's political goals and b) the individually self-interested motive in its most prosaic or venal forms. Social scientists studying rebellion or civil war have often compared and contrasted a) the rebellion's political 'grievance' with b) the rebels' individual motivation of 'greed' for pay, loot and other economic benefits. However, the debate about the relative importance of greed and grievance as motives for rebellion has apparently been replaced by a 'beyond greed and grievance' approach and a recognition that rebellion involves 'a complex mix of economic and political motivations' which appear to be 'concurrent' rather than mutually exclusive motivations.[13]

Such a concurrence or combination of collective-political and individual-economic motives would certainly answer the collective-action problems that were highlighted by Collier during the greed-and-grievance debate. He argued that a greed-motivated rebellion that secures sufficient funds or revenues to pay its recruits 'does not face any of the collective action problems of grievance-motivated rebellion', namely the well-known 'free rider' problem and the more obscure 'coordination' and 'time-consistency' problems.[14] These collective-action problems would also be solved by an insurgency that is motivated by collective-political grievance and also by the 'greed' to satisfy individual self-interested motives that may involve *non*-economic forms of 'pay off', such as social or moral (see Table 6.1). These non-economic pay offs are part of the wide range of positive and negative selective incentives that Olson identified long ago when explaining how collective-action problems were solved.

Collective-action problems

The collective-action problems were the focus of Olson's famous rational-choice analysis in *The Logic of Collective Action*. He argued that a large latent group will engage in collective action towards a collective good only if they are also motivated by selective incentives, that is, rewards or punishments that are separate from the collective good and operate selectively upon the group as individuals, such as offering them personal financial benefits for taking part in the collective action aimed at achieving the group's collective good.[15] Although the selective incentive is normally depicted as appealing to the individual's economic self-interest, it can also take the 'material' form of threatening to punish the individual by inflicting a personal injury. At its most extreme this coercive selective incentive will involve the selective use of violence as a punishment and deterrent.

Both these forms of selective incentive may motivate individuals to join an insurgent band or continue fighting for it. The use of threats to motivate insurgents to continue was mentioned in the previous section, and the next section will describe how the SL band resorted to using fear to recruit new members. Financial incentives to join and continue with the band have also been mentioned earlier and usually are an equivalent not of hiring mercenaries but of paying wages and salaries to professional soldiers, who are normally easier to recruit and retain in bad economic times than good.[16]

However, Olson's analysis also illustrates how *non*-material selective incentives can be used to overcome the collective-action problems involved in an insurgency.[17] 'Social sanctions and social rewards ... can distinguish among individuals: the recalcitrant individual can be ostracized, and the cooperative individual can be invited into the center of the charmed circle'.[18] He stipulated that these social selective incentives operated only in 'groups so small that the members can have face-to-face contact with one another', and therefore that large organisations using social incentives 'to mobilize a latent group interested in a collective good must be federations of smaller groups' in which the members have face-to-face contact.[19] Decades later Taylor pointed out that several studies had confirmed how social incentives were important 'in mobilizing people to contribute to collective action', with individuals contributing because they were 'asked and tacitly or overtly pressured by friends, colleagues, workmates, or co-members of the association's local branch or cell'.[20] He also noted that social incentives are particularly effective in groups that are relatively small and stable, as Olson would have predicted, but pointed to the isolated peasant-village community as exemplifying these characteristics.[21] In fact whenever the notion of social selective incentives has been applied to the study of revolutions, rebellions or insurgency, it has normally been in relation to rural communities and their ability to use social incentives to deliver recruits or support. Even students of rural rebellion who do not adopt an Olsonist approach may well acknowledge that it is very relevant in this context. For example, Petersen's analysis of the social mechanisms that assist the organisation of rebellion amongst rural communities acknowledged that the 'mechanisms of strong community – normative pressure and automatic social sanction, monitoring and effective threat – rely on relatively small numbers. Although the organization of rebellion is not necessarily a collective-action problem, much of the Olson logic of small groups and selective incentives applies'.[22]

Olson also pointed out that if small social groups are kinship or friendship groups, a psychological rather than social form of selective incentive might be used. He acknowledged that psychological incentives were important in such 'affective' groups as 'family and friendship groups' and that they could be analysed within the framework of his theory of collective action.[23] Although he was not interested in pursuing this application of his theory, family and friendship groups' psychological incentives seem to be an important factor in motivating individuals to engage in collective action on a much wider social stage. For example, studies of rebellions in Latin America, Eastern Europe and Ireland have shown 'that social network ties (especially friendship and kin ties) are the best predictors of joining a movement'.[24]

Furthermore, these are not the only form of affective motive that may be at work in the recruiting of insurgents. The seeking of revenge is a negative form of affective motive, and its antipathy may be directed at a large group rather than at a small group or an individual, but it is a powerful psychological-emotional motivation and has often been said to motivate some individuals to join an insurgency. For example, it has been argued that an important personal reason for individuals joining the Viet Cong in the 1960s was the desire to revenge civilians killed by misdirected bombing or artillery fire. 'When the desire for revenge was combined with appeals to patriotism, expelling foreigners, Viet Cong recruiters seem to have had a potent mixture with which to work'.[25] This seems a classic example of the combination of a collective-grievance motive (expelling foreigners) with a selective incentive (revenge) as the individual's self-interested motive. A similar combination is evident in Kramer's assertion that the desire for revenge as well as nationalist separatism has been the 'chief motivating force for all Chechen fighters' engaged in the insurgency against Russia.[26] However, in some parts of the world the revenge motive may be cultural as well as psychological, as when the recruitment of insurgents in 1980s Afghanistan was assisted by the desire to attain vengeance for family and friends killed by Soviet air attacks.[27] Considering that Afghan tribal society has a cultural tradition of seeking revenge for the killing of relatives and friends, any revenge motivation may have included a cultural element that bears comparison with Olson's notion of moral selective incentives.[28]

Olson not only identified a category of moral selective incentives but also distinguished them quite clearly from the social or psychological incentives that operate within social groups. He emphasised that 'adherence to a moral code that demands the sacrifices needed to obtain a collective good' is a form of selective incentive, and he argued that it exerts a coercive effect upon an individual through *self*-inflicted moral/psychological sanctions: 'the sense of guilt, or the destruction of self-esteem, that occurs when a person feels that he has forsaken his moral code'.[29] These moral selective incentives are therefore very much a matter of individual *self*-interest, in the sense of the interests of the individual's 'self'. But they are often not recognised by social scientists who are otherwise very aware of collective-action problems and how they are overcome by political organisers. A well-known example is Popkin's analysis of the role of collective-action problems in the communists' mobilisation of French-ruled Vietnamese peasants in the 1940s–50s. He clearly did not view 'ethics, altruism or conscience' or 'duty' as a form of selective incentive:

> When persons have decided to contribute on the grounds of ethics, altruism or conscience, a would-be leader need offer no selective incentives. ... During the August Revolution [of 1945 in Japanese-occupied Vietnam], there were many persons in Hanoi and Saigon who felt it their duty to fight for national salvation.[30]

He then noted the difference between these urban nationalists and a peasantry that usually needed selective incentives – specifically, material or social selective incentives – to motivate them to contribute to collective action of some kind. But clearly from an Olsonist perspective the distinction is between the moral selective incentives that motivated the urban nationalists and the very different forms of selective incentive that were needed to motivate the peasantry to 'fight for national salvation'.

On the other hand, other social scientists have built upon or paralleled Olson's insights about moral selective incentives.[31] His highlighting of the destruction of self-esteem that occurs when individuals forsake their moral code is similar to Calhoun's more recent formula that a self-sacrificing act can be a self-*saving* act in terms of personal identity, and Olson's rational-choice analysis of moral incentives is similar to Calhoun's analysis of the 'logic of honour', which depicts the individual as making a rational choice between a self-inflicted moral/psychological cost and an externally inflicted cost that may include physical torture and life-risking situations.[32] In choosing between the two kinds of cost the individual assesses 'which choice holds the greater pain', but choosing not to do 'the honourable thing' will endanger the individual's personal identity – thus choosing physical pain or danger will be a self-saving as well as self-sacrificing act.[33] The reality of these rational choices and assessments is dramatically revealed in the diary of a First World War US airman not long before his seemingly inevitable death:

> I only hope I can stick it out and not turn yellow. I've heard of men landing in Germany when they didn't have to. They'd be better off dead because they have to live with themselves the rest of their lives. ... I've got to keep on until I can quit honourably. All I'm fighting for now is my own self-respect.[34]

His moral incentive to continue fighting seems to be based upon a patriot's personal identification with the war being fought by his nation-state. And certainly when Calhoun used the notion of personal identity to describe rational choices about 'the self', he described personal identity in terms of an individual's identifying with a particular group or cause.[35] A famous historical example of this identification is the assassination of the Austrian Archduke Ferdinand in Sarajevo in 1914, which is commonly viewed as the catalyst for the outbreak of the First World War. The assassination was the work of a group of Bosnian-Serb conspirators whose identity-based sense of honour had been outraged by 'the Austrian annexation of Bosnia' – an 'insult to their national identity and a constant reminder of inferiority'.[36] Whether these assassins are categorised as expellers or separatists, their collective-action problems were apparently solved by a moral, not material, form of selective incentive that linked their ethnic-national identity to their own sense of personal identity.

The variety and mixture of motives

The wide range of selective incentives that are available in different circumstances is indicative of how wide a variety of motives influences individuals' joining and continuing with an insurgent band. In addition to the variety of selective incentives that apply to individual self-interest, there is the possible collective self-interest of different types of social groups and of course the ideological and national interests involved in political goals. Clearly the range of combinations, of various mixtures of motives, is immense and even within a particular insurgent band there will be a complex cluster of different motives. For example, in the Afghanistan of late 2003 the recruits for the reviving Taliban displayed a varied mixture of grievances and selective incentives: 'some needed the money that was being offered, while others were motivated by a strong desire to fight the Americans and their alleged "stooges". Many fought for a mixture of different reasons'.[37]

The mixture and variation within an insurgency is also evident in its recruitment methods, as was recently illustrated by Weinstein's account of Peru's Sendero Luminoso (Shining Path) – a largely rural maoist-model revolutionary insurgency. In the early 1980s its recruitment methods initially emphasised 'ideological appeals reinforced by ties to the rural peasant communities' but came to rely 'more heavily on fear and coercion', with political indoctrination of recruits 'sometimes commencing only after individuals had joined the movement out of fear'.[38] Initially the insurgents had been successful in selling the peasants the SL political goals and combining this political support with the use of what seems to have been a) moral selective incentives derived from ideological commitment and b) social selective incentives derived from the insurgents' ties with the local peasant communities. But the Peruvian state's growing military presence in these areas of the highlands had increased the risks that the peasant communities were taking in supporting the SL and therefore, to compensate for a decline in social selective incentives, the insurgents had to rely more heavily on coercion to overcome collective-action problems. In contrast, a regional SL splinter insurgency had been able to exploit economic selective incentives rather than resorting to the use of coercion. This splinter band had exploited the revenue opportunities provided by the drug trade in its coca-growing Huallaga Valley area and 'recruited almost entirely from a population of coca farmers, offering them an opportunity to enrich themselves further by carrying a gun'.[39] It had markedly adapted its recruitment methods to fit quite a different regional environment from that experienced by the main, more orthodox SL insurgency.

But even the orthodox SL insurgency had been forced to adapt its selective incentives to fit a new environment – namely, the new military-political environment created by the growing presence of the state's military forces. Recruitment methods for a particular insurgency therefore vary in different places and different times, and according to the changes in the insurgency's local environment that occur from place to place and time to time (discussed in Chapters 8 and 9). As Sageman concluded in the case of the jihadist insurgency, there can be only local and time-specific profiles, not an overall profile, of jihadists' recruitment.[40]

Yet there is a crucial distinction between changes in recruitment profiles or methods brought about by a) changes in the availability of particular forms of selective incentive and b) changes in the recruits' grievances and political goals. In the latter case the change may be related, for example, to 'world events occurring at the time and in the vicinity', as when 'some of the grievances motivating the jihad' in the 1990s included the 'presence of U.S. troops in the Arabian Peninsula, persecution of Palestinians, and former harsh sanctions against Iraqi children'.[41] So it is hardly surprising that the pool of possible participants in the jihad 'may increase or decrease according to socio-political events in the world'.[42]

In the 2000s that pool of possible recruits was being maintained by events in the Iraq, Afghanistan and other Western-linked conflicts. And there was a continuation of what Benjamin and Simon described as the 'jihad wired':

> Without the Internet, bin Laden still could have taken his jihad global – videotapes and compact discs were already at work before Netscape – but its growth would be at a comparative snail's pace. ... To grasp the impact the Internet is having, it is worth recalling how a different medium revolutionized attitudes in another time and place. In the America of the 1960s and 1970s, television brought the Vietnam War, as it was so often said, into the nation's living rooms.[43]

On the other hand, Sageman argues that 'the fear that vulnerable young Muslims may be recruited to the jihad through Internet messages is overblown', because joining the jihad requires both intense and lengthy 'face-to-face interactions'.[44] Such interactions occur only within a close-knit group of kin or friends whose affective-group selective incentives, their 'in-group love rather than out-group hate', overcome any collective-action problems involved in discussions and activities that strengthen the grievance motivation – and also provide a selective incentive for joining the jihadist insurgency.[45]

In contrast, the equivalent face-to-face interactions in SL-controlled areas of Peru had been the work of skilled SL indoctrinators. As mentioned earlier, during the peak of the SL insurgency its band could engage in face-to-face political indoctrination of potential recruits, before rather than after they had joined the band, and many other maoist-model insurgencies have enjoyed this advantage of being able to carry out indoctrination programmes without risk of intervention and attack by a state's forces. Another example of the advantages of maoist-model base areas is how the Vietnamese communist insurgency could guarantee potential recruits in rural base areas that the insurgency would take care of their families, such as giving a recruit's dependent grandmother an extra plot of land.[46] Such methods are not available to insurgencies recruiting from territories that are largely or fully under the control of a state's security forces. In these state-controlled areas the face-to-face indoctrination and the selective incentives must take a very different form.

Such differences in form are revealed by Sageman's biographical study of a hundred jihadists who had joined the jihad in the 1990s or early 2000s – and had joined from Western or Middle Eastern countries that were largely or fully under the control of a state's security forces. His definition of joining the jihad was 'the decision to go somewhere for training' as an insurgent, notably the training camps in Afghanistan, but only a minority of the arrivals were eventually 'invited to join' one of the jihadist bands; less than a third of Al Qaeda's trainees, for example, were invited to take part in a formal ceremony of acceptance and a loyalty oath.[47] Furthermore, in many cases the trainees had not been persuaded or invited by a jihadist to come to an Afghan camp or some other site for training and perhaps selection – instead they had recruited *themselves* to join the jihad and often recruited themselves as a *group*. As many as 75 per cent of them either had friendship or kinship bonds with a jihadist before joining or had joined with a group of friends or kin.[48]

In fact the trainees had often been self-recruited through male friendship groups that were 'clique networks' in social-network terminology but which Sageman preferred to label 'a bunch of guys'.[49] The guys' recruitment into the jihad involved the three stages of 1) joining a bunch-of-guys clique that 2) acquires a jihadist religious dimension through one of the relatively few jihadist-influenced mosques scattered throughout the Muslim world and parts of the Western world and then 3) finds a locally available 'broker' or 'bridge' jihadist who provides a link to a jihadist band's training activities.[50] This collective self-indoctrination and self-recruitment process would clearly have imposed severe social and psychological sanctions on any non-conformist. Not only would he have been subjected to the usual social pressures to conform to a group's norms but he would also have lost his friendship ties with the other 'guys' of his clique.[51]

Of course this three-stage process of joining the jihad has since been disrupted by the post-9/11 crackdown on training sites, 'broker' jihadists and jihadist-influenced mosques. Yet a similar process may have been involved in the creation of the self-starter, local-set microbands of post-2002 that were described in Chapter 5. Certainly the leader of such a band would have had powerful selective incentives available if his microband was based upon a bunch-of-guys clique. But by 2005 a different form of recruitment had emerged with the recruiting of volunteers to travel to Iraq and fight against the US and its allies. Like the 1980s recruitment of volunteers to travel to Afghanistan and fight the Soviet invaders, this process often involved a) an individual rather than a bunch of guys, b) the use of specialised recruiters or informal recruitment centres and c) one-stage recruitment of individuals to go and fight in the war – indeed to die fighting or in suicide attacks – rather than merely go somewhere for training and possible selection by an insurgent band.[52] In fact the recruitment of volunteers to fight in 1980s Afghanistan and 2000s Iraq seems quite similar to the top-down and talent-spotting process used by such insurgent bands as the Basque ETA insurgents in Spain.

ETA's specialised recruiters not only made the first approach to a prospective member but also gave him a series of minor tasks to perform as a test of his commitment and competence.[53] Furthermore, the selected recruit's decision to join ETA does not seem to have been affected by friendship or kinship ties and certainly was not the result of a group decision to join the insurgency:

> If ETA takes a long time in recruiting each member, it can also be said that the potential member takes his time in considering. Young men approached to become etarras typically resist joining for a long time before they cast their lot with the organization. Virtually every case study reports that the young man resisted the first invitations to join, sometimes for as long as a year and a half or two, before finally deciding to become a member.[54]

In these cases of individual recruitment some other selective incentives seemed to be operating, notably what Olson described as moral selective incentives. In particular, the individual's sense of personal identity and honour, with its accompanying self-esteem, might well be related to his Basque ethnic identity and the separatist cause espoused by ETA. Deciding to join the ETA insurgent band may be similar in its moral incentives and psychological coercion to the experience of a patriot deciding to 'join up' and fight for his nation-state.

Although it seems likely that some form of moral selective incentive has influenced the ETA recruit and the Iraq-bound jihadist, it is difficult to distinguish between this selective incentive and the collective grievance that is the basis of his insurgency's political goals. For example, members of the Madrid-bombing microband apparently had been affected by what they saw as the evils of the Iraq war and by watching videos of Muslims fighting in Afghanistan, Bosnia and Chechnya.[55] How these grievances can be transformed into a moral selective incentive is evident in the case of an insurgency with a very different ideological perspective – the Jewish Irgun band's insurgency against British rule in Palestine. In his memoirs Begin wrote that there 'are times when everything in you cries out: your very self-respect as a human being lies in your resistance to evil. We fight, therefore we are!'[56] And he gave eloquent testimony to how the 'cause' can motivate individuals not just to join an insurgency but to do extraordinary things on behalf of its political goals:

> But we did learn that an idea, after it has taken shape, mysteriously gives birth to the men who bring it to fruition. Recognizing this truth, I assert that if at any stage of the revolt all of us in the Irgun had been captured or killed, others would have taken our places and fought the oppressor until he was beaten. We learnt that the aim makes heroes of the weak, turns privates into officers.[57]

In this case the identification with a group and its cause means that the individual's 'self' can be expanded and made extraordinary – heroic and possessed of leadership qualities.

This is a positive aspect of moral incentives that was not explored by Olson or Calhoun but is particularly relevant when an individual's adherence to a moral code will have life-endangering or life-changing repercussions. Evidence of its effects can be found in Sageman's description of some dedicated jihadists whose 'awareness of their own readiness to transcend their self-interest fosters a special view of themselves' and whose 'conscious rejection of the free ride injects a semblance of virtue into the way they view themselves'.[58] Olson would doubtless have appreciated the paradox that the collective-action problem of free-riding can be overcome by the boost to self-esteem that comes from consciously rejecting the free ride.

A similar way in which the moral selective incentive may boost the individual's self-esteem is through the sense of pride in actually participating in the struggle. In her analysis of the Salvadoran insurgency, Wood emphasised this pride in participation and quoted the case of an elderly man who regarded his role in the insurgency as 'a great source of pride: to have stood up to it all'.[59] By analogy, it seems likely that participating in an insurgency strengthens some individuals' sense of personal identity, through identifying with a group and its cause. In fact the social psychologists' terror-management theory of group identification suggests that some individuals might be prepared to die for their nation, religion, culture or ethnic group because these individuals can overcome or 'manage' their fear of death by identifying with a group that will live on indefinitely and therefore is imbued with a kind of immortality.[60]

Obviously these considerations and forms of moral incentive apply to only a small proportion of insurgents. But as was emphasised earlier, the proportion is markedly higher when recruitment or self-recruitment occurs in countries or regions that are under the control of a state's security forces. For in these areas the insurgents are unable to use the more prosaic selective incentives, the coercion, material rewards and social rewards/sanctions, which are available to them when they have some military-political control over the territory from which they are drawing their recruits. The proportion of insurgents motivated by the less prosaic selective incentives would also be much higher in the crucial case of leaders and actual founders of a band or microband. It might be expected to be much higher, too, among those individuals showing the most self-sacrificing dedication in combat, notably the suicide attackers or those who display a virtually suicidal degree of commitment in more orthodox forms of combat.

The mention of suicide attackers also begs the question of religious motivation, as the suicide attacker seems to be commonly identified with the Islamic religion – and religion has not been included as one of the possible motives for modern rather than premodern insurgency. If a recruit's moral code is derived from a religion, it may be difficult to distinguish the moral selective incentive from a heavenly religious incentive, such as the heavenly rewards promised to Muslims who are killed fighting in a jihad. But as Elster argued when analysing suicide attackers' motivations, the desire for a 'religious hereafter', for attaining paradise, is better viewed as 'a form of consolation or bonus rather than a motivation'.[61] A different sort of scepticism was evident in O'Neill's argument that both religious and ideological types of belief are likely to be merely supplementary factors, and that it is the discontent arising from concrete grievances that makes individuals susceptible to religious or ideological rationalisations for resorting to violence.[62] Similarly, Sageman noted that a jihadist-inclined bunch of guys' increased devotion to religion 'may very well be a reflection' rather than cause of their increasing engagement in the jihad.[63]

Despite his dismissal of heavenly rewards as a motivation, Elster suggested that the desire for a *secular* hereafter, 'to transcend death by living on in the grateful or admiring memory of others', may be a powerful motivation for suicide attackers.[64] Some confirmation of this secular hereafter's existence can be found in Hopgood's account of how the deaths of the Tamil Tigers' many suicide attackers were remembered by an annual memorial day, during which a photograph of each hero or martyr was garlanded with flowers and had a 'flame of sacrifice' lit in front of it.[65] The Tamil Tigers were not Muslims, and also a sizeable proportion of the Middle Eastern examples of 1980s–90s suicide attacks were carried out by insurgents who apparently had a *secular* rather than Islamic orientation.[66] They actually predominated during the 1983–6 campaign of more than 30 suicide car-bombings in Lebanon (relatively few were carried out by the Shiite Islamist Hezbollah and most by pro-Syrian nationalists or leftists) but it is not clear whether these secular suicide attackers were motivated by aspirations for a secular hereafter or some other motives.[67] However, there seems little doubt that history's most influential insurgent action – the shooting of Archduke Ferdinand in Sarajevo in 1914 – was partly motivated by the desire for secular immortality. For the man who pulled the trigger:

> willingly exchanged a finite lifetime for boundless renown. In prison he wrote down a quotation from an earlier Bosnian martyr, 'He who wants to live, has to die. He who is ready to die, will live forever'.[68]

Such a pay off is perhaps the opposite extreme to the common, coercive and economic forms of selective incentive and certainly has already motivated an individual to change the course of history.

7
The Means of Insurgency

The means variable of the insurgency calculus has already been partly examined in several chapters of Part I, for this variable includes a) the method of insurgency as well as b) the capability for engaging in insurgency. The insurgents' method includes the mode of operations, the operational format and of course the strategy being pursued, all of which were discussed in several earlier chapters. So in this chapter the focus is on capability rather than method, and particularly on the personnel, information and equipment components of insurgents' military capability.

However, the personnel component, specifically the quality of personnel, also overlaps with the motive variable. For example, a band's capability to carry out suicide attacks involves the use of highly motivated volunteers. This requirement was almost farcically illustrated by the attempts of the Colombian FARC to find a volunteer for its planned suicide operation of crashing a plane into the presidential palace. Although it was reportedly offering to pay two million US dollars for someone to carry out the attack, no volunteers came forward and the operation had to be replaced by a series of bombings timed to coincide with the inauguration of the new President.[1] Considering the importance of motivation in this and so many other cases of insurgent activity, the chapter begins with a discussion of motivation as capability before going on to examine the personnel, information and equipment factors.

Finally, the capability aspect also includes the quality of the band's leadership. In particular, it includes the band leaders' intellectual ability to set a direction successfully and to make adaptive changes in direction, notably by changing the band's method. The leaders of bands therefore require the situational awareness that will enable them to assess not only capability and method but also the various forms of opportunity offered by their local environment, which will be discussed in the next two chapters.

Motivation as capability

The motivation of insurgents is clearly an important part of their military capability, just as it has always been for any military unit that faces the prospect of killing and being killed. In the 18th century Prussia's King Frederick the Great famously remarked that his fiercely disciplined conscript army should fear their officers more than they feared the enemy, but he also acknowledged that their officers were in turn motivated by the aristocracy's code of honour: 'the one factor which can make men march into the cannon that are trained at them is honour'.[2] With this combination of motives (which social scientists would now describe as coercive and moral selective incentives), Frederick was able to deploy a large army that suffered heavy casualties and several defeats in defence of an 'artificial country' which could not exploit patriotism or nationalism to motivate its soldiers. A hundred years later the armed resistance of indigenous peoples to 19th century European imperialism produced an even higher level of motivation that again was not linked to the modern or Western concepts of patriotism and nationalism. The 'fanaticism' of these peoples' resistance was cited by British authorities as justification for using the Dum-Dum (internationally banned) expanding bullet in colonial wars against 'fanatical natives', who were described as 'peculiarly hostile and resolute, besotted with fanaticism'.[3]

Modern insurgents' motivation may be less fanatical but is no less important. For example, the US Marines holding the city of Ramadi in Iraq in 2004:

> were impressed by the dedication and motivation of their enemies in this area. A Marine sergeant, who was later to be killed there, said: 'They were young just like me. Fighting for something different, something they believe in. And that's the worst kind of enemy.'[4]

Furthermore, some modern insurgents are as 'fanatical' as some pre-modern insurgents and have been sufficiently motivated to carry out suicide attacks, which therefore means that their bands have the capability to use what is sometimes called 'the poor man's smart bomb'.

The suicide attack may have been overshadowed in recent years by the roadside bombs used by Iraqi, Afghan and Chechen insurgents but it remains a very effective technique for attacking human beings who are not protected by armour plate or strong security precautions. An obvious reason for the effectiveness of the suicide attack is its ability to deliver an explosive device (even if only a relatively small device concealed on the attacker's person) to its target by infiltrating heavily guarded targets and by making last-minute adjustments, such as when setting off the device at the right moment or when ramming a vehicle into a target.[5] Even before 9/11, suicide attacks had proven much more deadly than insurgents' conventional attacks on 'soft' targets. Although there had been fewer than 200 suicide attacks since 1980, only a small percentage of all 'terrorist attacks', they had nonetheless killed nearly half of the more than 3,000 victims of such attacks during this period.[6]

As Pape and others have pointed out, there is an obvious similarity with the effectiveness of suicide attacks carried out in warfare between states, notably the Japanese kamikaze airplane attacks on Allied ships in 1944–5. They were 'considerably more effective' than equivalent conventional attacks, with almost 12 per cent of the nearly 4,000 'special attack' missions achieving a hit on enemy ships.[7] A less obvious comparison is with the Japanese use of suicide human-bomb attacks against tanks. In its 1945 defence of Manchuria against the Soviet Union, the Japanese army lacked 'anti-tank capability' and therefore resorted to having soldiers use explosive charges against enemy tanks in human-bomb fashion: 'Most Soviet tank losses came at the hand of these suicide attacks'.[8] This is very similar to the LTTE Tamil Tigers' use of a specially trained and dedicated unit of suicide attackers, the Black Tigers, who were used primarily for attacking conventional military targets, such as being employed as a very accurate or 'guided' missile-like delivery mechanism for destroying enemy fortifications.[9]

But if an insurgent band's members lack a suicidal level of motivation, then obviously the band will lack the capability to launch suicide attacks. And in fact not many insurgent bands have such motivated personnel. A quite recent analysis of why so few insurgent bands have used suicide attacks points out that the ability to do so depends upon individuals' willingness to volunteer for suicide missions as well as upon the organisation's willingness to use this form of attack.[10] As was described earlier, such a well-established and well-known insurgent band as the Revolutionary Armed Forces of Colombia (FARC) may face an embarrassing shortage of sufficiently motivated personnel.

FARC's problems with insufficiently motivated personnel was also illustrated by the fact that its refusal to pay its soldiers the equivalent of a salary was a major reason for desertion from its forces.[11] As might have been expected, even greater problems were faced by the Sendero

Luminoso splinter group in the Huallaga Valley that relied almost entirely on drug-related economic incentives; when the Peruvian government introduced an amnesty law, it is estimated:

> that more than five thousand Senderistas and civilian supporters gave themselves up to the police in the Upper Huallaga Valley. Given that Sendero Luminoso had at most ten thousand soldiers and tens of thousands of supporters nationwide in 1990, the capitulation of five thousand in Huallaga alone was a significant blow.[12]

Even Che Guevara was faced with the problem, in Bolivia in 1967, that 'recruits enticed by money or promises were more likely to desert after their first brush with combat'.[13]

Although it is unlikely that insurgent recruits will be merely mercenaries, in some insurgencies a large or significant proportion of the recruits will have joined 'for the money' or some other economic benefit – which will affect the capability of the insurgency in more ways than the desertion rate. For example, the economic motivation that seems very prevalent among the African roving bands and hybrid bands has also had significant effects on their capability to pursue particular strategies. Ballentine has suggested that the influence of the economic motive, focused on lootable resources, may be a reason why some African rebel groups have failed 'to generate a coherent political program or to attract sustained popular support' and may also be a reason for the 'relatively higher level of indiscriminate violence perpetrated against civilians by these groups'.[14] Such indiscriminate violence against civilians would in turn be a reason to expect that these bands have little capability of pursuing a propaganda or provocation strategy, as was discussed in Chapter 4. Furthermore, their capability of pursuing a pressure or preparation strategy may have been militarily handicapped by their need to shift to semi-state warfare, however prematurely in strategic terms, merely in order to acquire the resource-rich territory that they need to recruit and retain economically motivated insurgents.[15]

Insurgents' military capability

The wider or overall military dimensions of capability go beyond motivation and include a wider range of factors. They can be summed up by the formula Personnel, Information and Equipment (PIE), which also equates in administrative language to human resources, software and hardware. From the PIE perspective, motivation is merely a sub-category of one of the two major subdivisions of Personnel, which is subdivided into 1) quantity and 2) quality in terms of a) motivation, b) training and c) adaptability.

The quantity or sheer number of personnel available to a band is the most obvious factor determining its capability and can have a huge influence upon the band leadership's choice of methods, especially of what strategy or strategies to adopt. Numbers are not so important to the success of propaganda and provocation strategies but, as was discussed in Chapter 4, these are only supplementary or intermediate strategies and cannot themselves achieve the insurgents' political goals. On the other hand, the two strategies that can achieve victory, preparation and pressure, have always required a significant and often very large number of personnel.

A striking example of the effect of insufficient numbers is the 1970s era of urban revolutionary bands in Western Europe, notably the German Baader-Meinhof group and the Italian Red Brigades. Townshend has described them as 'groupuscules' or 'tiny, fissiparous radical activist groups', whose robberies, kidnappings and other attacks had a 'corrosive impact' upon 'public morale and civic confidence in Germany, Italy and beyond'.[16] These small bands and microbands of course did not seriously consider a maoist-model strategy and instead adopted the intermediate – and foredoomed – strategy of provoking the state into a level of repression that would radicalise public opinion.

A less extreme example of insufficient numbers is that of the resurgent Taliban in 2007. The Taliban still had only about 12,000 fighters (including part-timers) and therefore could not seriously consider a maoist-model strategy of eventually shifting to semistate warfare and defeating on the battlefield an Afghan army that already numbered 35,000 troops and was planned to number eventually 70,000.[17] Clearly even if their pressure strategy had succeeded in expelling the Western forces, the Taliban were hardly in a position to repeat the 1991 success of the much more numerous mujaheddin in using semistate warfare to 'march on the capital'.

A pressure strategy is less dependent on numbers than a maoist-model preparation strategy but nonetheless needs sufficient numbers to inflict a politically significant amount of hurt. For example, the Provisional IRA of the early 1990s had 'at most 500 actual fighters', which hardly compared favourably with the '3,000–5,000 active IRA men who took part in the struggle' of 1919–21.[18] Begin's Irgun was probably the smallest band to have successfully pursued a pressure strategy, for it never had more than a few dozen full-time operatives and for some time had only hundreds rather than thousands of part-time members.[19] But the Irgun was fighting an expeller, not separatist, insurgency against a British government that was vulnerable to political pressure, international as well as domestic, to end its colonial rule over Palestine. Furthermore, Begin's memoirs implied that the Irgun supplemented its pressure strategy with a state-weakening propaganda strategy. Realising that British colonial officials normally preferred to base their rule on prestige rather than on force, the Irgun believed that 'if we could succeed in destroying the government's prestige', then 'the removal of its rule would follow automatically' – and they were aided by the fact that 'when a colonial regime lives by the legend of its omnipotence', every insurgent attack that it has failed to prevent 'is a blow at its standing'.[20] The small size of the Irgun band did not prevent it from pursuing a successful propaganda strategy that supplemented its pressure strategy against a politically vulnerable foreign state.

The quality of a band's personnel can be almost as important as their quantity in providing opportunities for using particular strategies and even tactics. Motivation was discussed in detail in the previous section but the training aspect of personnel quality has been only briefly mentioned, largely in relation to Mao's concern with establishing base areas in which a regular army could be trained for semistate warfare. Obviously without such training facilities there is no hope of pursuing a maoist-model strategy – and cyberspace is no substitute for training grounds when it comes to preparing a regular army.

Furthermore, even the *ir*regular warfare of insurgent hit-and-run attacks requires some training grounds if the attacks will involve actual combat. Although the arrival of the Internet has enabled insurgents to be taught in their own homes how to make bombs and disable tanks, this distance-learning decentralisation is no substitute for hands-on combat training or indeed instruction in carrying out any complicated and coordinated operation.[21] The US troops' superior training was no less important than their superior weapons in giving them a crucial edge in combat against Iraqi insurgents, while in 1944 the French Resistance insurgents' lack of combat and weapons training had a serious effect on their ability to fight German troops.[22] Similarly, jihadists lacking in proper training have had difficulty carrying out complicated and coordinated bombing attacks, as in the case of the 2003 Casablanca bombings. Most of the jihadists involved in this attack had been locally and poorly trained, had solved their bomb-making problems by finding a formula on the Internet, and had then proved incapable of carrying out their sophisticated operational plan – four of the planned five simultaneous bombings were more or less failures.[23] Even if it is true that Internet training has 'gone a long way toward eliminating the need for terrorist training camps and even for sanctuaries', this cyber-training also clearly has its limits; the Internet's greatest effect on insurgents' capability may have instead been its boosting of their adaptability, as the sheer range of experts and experience available to Internet users has increased the speed at which improvements and innovations spread around the world.[24]

Adaptability is itself another aspect of the personnel-quality factor. In fact military experts have recently suggested that 'the ability to adapt is probably most useful to any military organization and most characteristic of successful ones'.[25] A contemporary example of how it can contribute to the survival and success of an insurgency is the way that the Taliban have so rapidly and effectively adapted to the introduction of Western military forces and techniques:

> Today's enemy represents the third generation of Taliban that we have encountered (the first generation was the irregular conventional force ... which we destroyed in 2001–2002; the second generation was the embryonic guerrilla force that emerged from the remnants of the first in the Pakistan safe haven in 2003–2004 and began raiding into Afghanistan in earnest in 2005). In contrast to previous generations, today's Taliban are an extremely proficient, well-organized, and well-equipped insurgent force.[26]

Insurgencies typically require not only their senior leaders but also quite lowly personnel to show qualities of initiative and therefore of adaptability. The jihadist insurgency, in particular, 'prominently features local initiative and decentralized decision-making'.[27] Such a long-range, complicated operation as the 9/11 attacks required operatives with the adaptability to come up with their own solutions to unforeseen and complex problems – and these operatives are hard to come by. When Atta and three other men who would become key 9/11 operatives travelled from Germany to Afghanistan in 1999 for jihadist training, they were thankfully selected by Al Qaeda for an operation 'which had been in the planning stages for years' but had been missing 'volunteers familiar with Western countries, able to solve complex problems and work independently'.[28] If such adaptable personnel had not become available, the operation would presumably have had to be cancelled to prevent an embarrassing failure. In this case the band's capability to carry out a successful provocation strategy was dependent upon the availability of personnel with the required adaptability for the task.

Adaptability can also be an important feature of the personnel's tactical techniques. For example, the Viet Cong's continual self-appraisal and self-criticism helped them keep pace with the US forces' introduction of new technology, such as the use of helicopters and air strikes.[29] The Viet Cong 'soon adapted to the changed conditions, which no longer permitted them to lay siege to a fortified village or post; instead, they learned to batter its defenses to pieces in one hammer blow early enough in the night to be able to crush its resistance before daybreak'.[30] The Iraqi insurgency has seen combat adaptability go beyond learning the hard way how best to counter the US tactics; it has also included learning how best to trigger (by wire, by radio signal, by jamming-triggered continuous signal and ultimately by encrypted continuous signal) the roadside bombs that by 2005 were accounting for more than half of US combat deaths.[31] Western experts were impressed by the sheer number of tactical and bomb-making adaptations as well as by the speed with which the insurgents evolved and diffused these adaptations, which included the use of explosively-formed-projectile technology to destroy heavily armoured vehicles.[32] A similar adaptability was shown by Chechen separatist insurgents in their use of roadside bombs in 2003–4, notably their development of 'daisy chain', inter-linked multiple bombs that could destroy the most heavily armoured vehicles.[33]

Despite the importance of adaptability and the other aspects of the Personnel factor, in some situations the Information factor can be almost as important for the success of an insurgency. Information or the ambiguous term 'intelligence' is seen by military experts as providing a) warning of what the enemy is going to do and b) knowledge of the enemy's forces: 'who, what and where'.[34] The quality of information is of course as important as the quantity, particularly as the latter can lead to processing or interpretation problems for under-staffed insurgent bands. The importance of high-quality information is illustrated by the IRA's famous successes against the British in 1919–20. Collins developed 'an extraordinary network of agents' that even extended into the British headquarters in Dublin Castle, where his agents included police detectives, a senior civil servant and a coding clerk.[35] However, it has been argued that the British closed the 'intelligence gap' with the IRA so quickly in 1920–1 that Collins's legendary intelligence system now seems to have indeed been more a matter of legend or myth than reality.[36] And an Irish historian has noted that one of the pressures pushing Collins towards some compromise with the British government in 1921 was the change in the intelligence factor: 'the British had regained the initiative in this field'.[37]

The Equipment factor obviously makes a huge contribution to insurgents' capability of inflicting a high casualty rate and thereby exerting indirect political pressure upon the enemy state. In fact the 1950s Mau Mau insurgency against British colonial rule in Kenya illustrates how a pressure strategy can be foredoomed by a lack of arms, ammunition and explosives. Although at one stage over 15,000 Mau Mau were hiding out in the forests, less than a tenth of them had firearms and the rest were 'equipped' with only spears, bows or machetes.[38] The insurgents had begun stealing guns and ammunition in 1949 in preparation for their revolt but even in the peak year of 1953 they acquired fewer than 700 firearms from civilians, soldiers and the security services.[39] Although they took to manufacturing their own firearms from water piping, carved wood and tyre inner-tubing, the Mau Mau could not manufacture ammunition or explosives and so were very short of ammunition and wholly lacking in explosive devices.[40] If they had been equipped with the AK47s, RPGs, and IEDs available to the Iraqi insurgents 50 years later, the Mau Mau would have been able to exert substantial political pressure through their military power to hurt. They could have inflicted severe casualties on the British conscript soldiers who were guarding the colony's vulnerable supply routes and were carrying out such vulnerable counterinsurgency operations as the massive Operation Anvil, which rounded up thousands of suspected Mau Mau living in the city of Nairobi.

The equipment factor can change even more quickly than the information factor, particularly with the introduction of new technology. The most famous example is the Afghan anti-Soviet insurgents' acquisition in 1986 of the Stinger shoulder-fired surface-to-air missile. The new missile 'had an immediate military impact' and 'clearly represented an enormous qualitative improvement in the rebels' air-defense capability'.[41] On the other hand, it may not have had such an impact upon the insurgents' capability in terms of their pressure-based strategy or even their merely operational and tactical methods. The Soviet armed forces rapidly introduced countermeasures that reduced the Stinger's operational and tactical impact and, strategically, the Soviet Union's new leader, Gorbachev, had already decided that he needed a political settlement with the insurgents in order to improve the Soviet Union's political and economic relations with the West.[42] So the Afghan insurgents had already succeeded in carrying out a pressure strategy, if only by prolonging the war sufficiently to reap the benefits of a new, reformist leadership generation coming to power in the Soviet Union.

A better example of how change in the equipment factor can affect capability is Hezbollah's use of wire-guided antitank missiles against

Israeli fortifications in the cross-border Security Zone area of southern Lebanon. Introduced in January 2000, the new missiles were responsible for most of the Israeli fatalities in the following months and therefore may have had a major effect on the political pressure from Israeli society that led Prime Minister Barak to announce in May 2000 that Israeli troops would withdraw to the international border with Lebanon.[43] The fact that the insurgents now had a missile that could be guided through the observation ports of previously impregnable fortifications hardly seems a major shift in the military balance, let alone the political calculus of withdrawal. But the change came after several years of successful attacks on Israel's and its SLA ally's forces in southern Lebanon. There had been a change in the ratio of Hezbollah to Israeli/SLA casualties 'from more than 5:1 in 1995 to less than 2:1' through such techniques and tactics as 'cleverly camouflaged remote control roadside bombs'.[44] The trend or momentum of the war was therefore going so badly for the Israeli government that one more step in that direction, such as a new vulnerability being revealed by the deployment of a new weapon, would markedly increase the insurgents' capability to pursue a pressure strategy. However, the increase in capability was partly due to a threshold or tipping-point effect that had been produced by earlier changes in the military and political environment – by a weakening of the enemy's military and political position that had increased the *opportunity* for the strategy and the expeller insurgency to be successful.

Leadership – awareness and adaptability

Before examining the opportunity variable, though, there is still a key element of insurgent capability that needs to be recognised and explored – namely leadership. Chaliand has argued that in the case of insurgency 'the calibre of the leader is as important as, and in some circumstances more important than, the quality of the troops' and that a 'leader's strategic, political and military savvy has consistently been as crucial as popular support. It is not enough to have a following: one must also know *how to get to where one wants to go*'.[45] O'Neill confirmed that 'a serious insurgent threat will not emerge unless there are determined leaders who have the requisite skills to exploit potential opportunities'.[46] And even the Marxist, class-oriented Debray acknowledged 'the irreplaceable value of the leader, especially in the first stage' of the insurgency.[47]

However, knowing 'how to get where one wants to go' and skilfully exploiting potential opportunities require a very broad 'situational awareness'. Effective leaders of insurgent bands are aware of not only all aspects of the local environment but also the motivation and other capability of the band for using any particular method. The need for an awareness of the situation within the band was recognised by the Tupamaros when they developed a sophisticated environmental theory of the conditions (*coyunturas*) for a successful insurgency in 1960s Uruguay. They believed that 'the effectiveness of the guerrilla movement is partly a function of the ability of its leaders to recognize proper *coyunturas*'.[48] And these leader-recognised conditions included the capability of the band to exploit a social opportunity:

> [For] the social movement (in this case, the guerrillas) might not be ready to take advantage of such an opportunity. As an example, urban warfare may be viewed as a viable instrument of political violence in a highly urbanized society; but the lack of manpower, leadership, or arms and ammunition can make it impossible for a social movement to undertake such a course of action. ... For the Tupamaros, then, the *choice of every strategy and tactic* was the result of a careful, rational analysis of the present and potential strength of the *guerrilla group* as well as of the general conditions and political climate of Uruguayan society'.[49]

Insurgent leaders' situational awareness therefore involves three different types of assessment (see Table 7.1). It requires assessing the band's 1) motivation and other capability for using 2) an appropriate method 3) to exploit a perceived opportunity offered by the local environment. Only then will the leaders be able to perform successfully the basic

Table 7.1 Insurgent Leaders' Situational Awareness – Three Types of Assessment

1) Capability for an appropriate	2) Method to exploit the	3) Opportunity
a) personnel (including motives)	a) strategy	a) in the physical, rural/urban environment
b) information	b) format	b) in military and political environments
c) equipment	c) mode	c) in the social-political environment

leadership role – setting a direction and changing that direction when the situation requires it.

The key changes in direction are the adaptive changes in method, which are made by imitating a successful model or by making a local adaptation to the nature of the band or its local environment. The difference between imitating a model and making a local adaptation is illustrated by two historical cases described in a text on terrorism. Imitating a successful foreign model is implied by the remark that the leaders of 'the Irgun studied the tactics of the Irish Republican Army's Michael Collins. They tried to incorporate Collins's methods in the Jewish campaign'.[50] A local adaptation is implied by the remark that the PLO adopted terrorism because 'It was the only tactical option available'.[51] The difference between imitation and local adaptation is also described, by analysts of military failure, as the difference between a failure to learn and a failure to adapt – and it is probably in preventing such avoidable failures of learning and adapting that the band leaders perform their most important role.[52]

An excellent example of this role was provided more than 50 years ago by General Grivas, the leader of the Cypriot EOKA insurgents fighting to expel British colonial rule from Cyprus. The insurgency had initially adopted a largely rural bandit-like format but was almost destroyed in 1957 by the British army's counterinsurgency campaign in the island's mountainous regions. Grivas therefore set a new direction that adapted to the local environment's greater opportunity to use an urban gangster-like rather than rural bandit-like format. The change of method from rural to urban format forced Britain to deploy some 40,000 troops against the insurgents, and although fewer than a hundred troops were killed, the cost was too high for the British government to continue with colonial rule.[53] It was a classic example not only of an insurgent leader's adaptability but also of how the leader's failure to adapt would have led to the failure of the band and its insurgency.

However, the most impressive example of a leader's adaptability involved the setting of an innovative direction for a *new* insurgency – the expelling of the US forces that in 1993 were assisting the UN effort in clan/warlord-ruled Somalia. Somali warlord General Aidid's expelling of US forces required a series of adaptive changes in the methods that his localist, clan-based band had used against previous foes, namely other clan-based bands and the Somali state that they had destroyed in 1990. Aidid even sought to modernise his band's stateless warfare by imitating the nationalism of anti-colonial expeller insurgencies, such as having Radio Mogadishu broadcast a propaganda campaign depicting the UN and US as foreign occupiers of the country.[54] Strategically, he learnt from the defeat in 1920 of a Somali nationalist-Islamist leader against British and Ethiopian imperialism, Sheikh Abdulle Hassan. General Aidid had criticised him for shifting from guerrilla warfare to a more conventional warfare that enabled the British to exploit their superior firepower and biplane bombers.[55] So Aidid did not make the mistake of adopting a maoist-model preparation strategy and establishing semi-state base areas that would have been vulnerable to US aerial firepower; he instead adopted a pressure strategy that maintained the insurgents' advantage of being an enemy who is 'everywhere and nowhere' while inflicting sufficient casualties to increase the political pressure on the US President to withdraw his troops. Furthermore, Aidid reinforced a recent local adaptation by Somali clan-based bands, who were shifting from a rural bandit-like to an urban gangster-like format – which also abandoned traditional restraints on warfare and no longer distinguished between warriors and non-combatants:

> He was a skilled military man who applied the traditional Somali way of war to the urban setting. Small militia units – guerrillas – adapted to the urban terrain for close-quarter combat. ... [And] Aidid's warriors found yet another tactical use for women and children. They became human shields from behind which these gunmen attacked the soldiers of Task Force Ranger.[56]

However, there was no need for Aidid to adopt a terrorist mode of operation, because his pressure strategy sought to inflict hurt through military casualties rather than by attacking US civilians in the region or in their homeland. After the 3 October battle that produced over a hundred dead or wounded US soldiers, the political pressure on the President was sufficient to produce a US withdrawal – and a dramatic example of how effective the pressure strategy can be when the insurgents are directed by an effective leader.

8
Opportunity and Environments

These final two chapters, Chapters 8 and 9, are concerned with the opportunity variable of the insurgency calculus and with applying an environmental approach that views insurgents' opportunity as a product of their local environment. The physical, political, military and social aspects of a band's local environment can each be viewed as a separate environment. Each of them also presents a particular type of opportunity, such as a political or military opportunity, which when added together constitute the overall opportunity variable of the insurgency calculus. This chapter will examine the physical, political and military environments, especially the very local military-political environments known as 'operational zones' and their relationship to insurgents' support bases; the next chapter will focus on the social-political environment that may or may not provide the political support needed to expand and/or strengthen insurgents' support bases.

A feature of the opportunity variable is that it often has the effect of negating or counterbalancing the motive and means variables by being too weak to give the insurgents any chance of achieving a significant degree of success. To take a recent and prominent example, the Iraqi insurgents of 2003–6 probably had the strongest capability of any post-Vietnam insurgents, thanks to a) the number of trained personnel provided by the disbanded armed services and security services of the former regime; b) the amount of equipment available in the numerous ammunition dumps and weapons caches left behind by the former regime; and c) the adaptability shown by the insurgents at strategic, operational and tactical levels.[1] Yet they were unable to expel the US military or to bring about a partial withdrawal of US forces because from 2007 onwards their military environment – the military measures being taken to counter their insurgency – gave them no opportunity, not even a weak opportunity, to be even a partially successful expeller insurgency. Another example, in this case more general and hypothetical, is that any revolutionary insurgency is bound to fail in a democracy because the political environment gives these revolutionaries no opportunity of achieving any success at all. In the famous words of Che Guevara: 'Where a government has come into power through some form of popular vote, fraudulent or not, and maintains at least an appearance of constitutional legality, the guerrilla outbreak cannot be promoted, since the possibilities of peaceful struggle have not yet been exhausted'.[2]

However, there are also many examples of the local environment providing a strong opportunity for insurgents or at least sufficient opportunity not to negate or counterbalance their strong motivation and/or effective means. For example, a democratic political environment does not negate the possibility of a separatist rather revolutionary insurgency achieving some degree of success. And the local physical environment may actually provide a strong opportunity for insurgency through such advantages as extensive mountainous terrain or proximity to another country offering sanctuary or a safe haven. In fact there is often a need for a net assessment to be made within the opportunity variable – adding together the strength and weakness of the various types of opportunity – before any lateral net assessment can be made of the opportunity variable versus the other two variables of the insurgent calculus.

The other complication with the opportunity variable arises from the fundamental, 'ontological' fact of life that local environments change in time as well as place. Mao's distinctively environmental conception of war included the historical changes that occur in any local environment as it changes from one era, generation, decade or year to the next:

> Thus the different laws for directing different wars are determined by the different circumstances of those wars – differences in their *time*, place and nature. ... In studying the laws for directing wars that occur at *different historical stages*, that differ in nature and that are waged in different places and by different nations, we must fix our attention on the characteristics and development of each, and must oppose a mechanical approach to the problem of war.[3]

For example, there was a historical or temporal change in the local military environment of Iraqi insurgents in 2007 when the US corrected its earlier counterinsurgency failings and thereby substantially weakened the insurgents' military and overall opportunity for success.

By the same token, the fact that local environments change from place to place as well as from time to time implies that a similar counterinsurgency effort may not have the same effect on the Taliban's military environment in Afghanistan. The new counterinsurgency effort may merely reduce rather remove the Taliban's degree of success, not only because the Taliban differ in method and capability from the Iraqi insurgents but also because the non-military environmental factors are so different in Afghanistan that they counterbalance whatever changes may occur in the military environment. Mao clearly envisaged the local environment as changing from one geographical place to another, if only in the sense of changing from one country to another. 'As for the factor of place, since each country or nation, especially a large country or nation, has its own characteristics, the laws of war for each country or nation also have their own characteristics, and here, too, those applying to one cannot be mechanically transferred to the other'.[4]

The physical environment – rural and urban

The most obvious application of an environmental approach is to the physical variations in geographical 'place' that insurgents experience – the differences in terrain and geographical location. Insurgents operating in rural areas are obviously greatly affected by this local physical environment and the opportunities that it provides or fails to provide. In his history of warfare Keegan noted that there are physical influences on war-making that 'have always lain beyond the will or power of man to command' and that 'such factors – weather, climate, seasons, terrain, vegetation – always affect, often inhibit and sometimes altogether prohibit the operations of war'.[5] These varying physical factors are more important to guerrillas than to soldiers fighting a conventional war. As Galula noted, if the guerrilla 'cannot get any help from geography, he may well be condemned to failure before he starts'.[6] Mao specified that mountainous regions offer the best conditions for guerrillas, so long as they have enough room to manoeuvre.[7] And he was vindicated in the 1980s by the Afghan expeller insurgents' remarkable success against the Soviet superpower and the Eritrean separatist insurgents' successful shift to semistate warfare against the Ethiopian state. The Eritreans had relied greatly on their region's mountainous terrain as the springboard for their successful maoist-model preparation strategy, while the Afghan expellers had exploited Afghanistan's mountainous terrain and huge territorial space to outlast their superpower enemy and thereby win a great victory for the pressure strategy. However, O'Neill's analysis of insurgents' physical environment mentioned not only mountains but also 'jungles, swamps, forests and the like' as being vital for rural insurgents, because such terrain 'hinders movement by government troops and provides inaccessible hideouts for the guerrillas' main bases'.[8]

Furthermore, Galula included the country's *location* as one of the geographical factors that can help the insurgent get started. For as he pointed out, long inland borders with sympathetic countries will favour insurgents because they offer the prospect of sanctuaries or safe havens as well as external sources of supplies, and this was illustrated in the 1980s and 2000s by the benefits that Afghan expeller insurgents won from their long inland border with sympathetic Pakistan.[9] Galula even included a hypothetical example of the conditions that best favoured a rural revolutionary:

> To sum up, the ideal situation for the insurgent would be a land-locked country shaped like a blunt-tipped star, with jungle-covered mountains along the borders and scattered swamps in the plains, in a temperate zone with a large and dispersed rural population and a primitive economy.[10]

Yet it is surprisingly difficult to find reliable or uncontested statistical evidence about the influence of the physical environment as an opportunity factor. On the one hand, a cross-national statistical analysis of 1940s–90s civil wars and large insurgencies seemed to confirm the importance of the terrain factor. Among the variables the study used was one that measured the presence of 'rough terrain' likely to be suitable for guerrilla warfare, and a significant relationship was found between mountainous terrain and higher rates of insurgency and civil war.[11] On the other hand, a more recent study found that although mountainous terrain is 'a key variable in theories of civil war that emphasize the opportunity structure for rebellion', it is also a variable that loses significance when changes are made to the coding rules – when different criteria are used for deciding which historical conflicts should be included in the statistical analysis.[12] As insurgencies have varied so much historically and geographically as well as in type and strategy, perhaps it is not surprising that they have confounded the statisticians as well as all the other kinds of experts who have studied insurgent stateless warfare.

Another complication is that some of the beneficial effects of the physical environment can be nullified by a counterinsurgent state that has sufficient resources or foreign aid. Obviously a state should be able to improve its transport network – and thereby improve its troops' mobility and access to rough terrain – by investing heavily in this militarily crucial part of its economic infrastructure. And as O'Neill pointed out, the new aerial mobility, especially the use of helicopters, has markedly reduced the effects of not only terrain but also territorial space.[13] Yet there is still some truth to Mao's argument that guerrillas' opportunity is also related to the size of a country and the extent of its state's control over this territorial space. For Ballentine noted quite recently that the limited reach of Third World states' authority in remote border areas and other peripheral locations means that 'the opportunity for rebellion' is also shaped by the 'spatial aspect of state weakness', which can provide insurgents with the equivalent of mountainous terrain.[14]

As Guevara noted, bands which choose inaccessible and sparsely populated locations thereby run the risk of becoming secure but ineffectual revolutionaries.[15] However, insurgents who are separatists or expellers rather than revolutionaries may not be running the risk of ineffectiveness if they choose a secure but inaccessible and sparsely populated location. The obvious example is the way in which the Al Qaeda headquarters was and is protected by its remote location and mountainous terrain – in Afghanistan before 9/11 and since then apparently in the Afghanistan-Pakistan border region. Thanks to the new techniques and technology used by third-phase insurgents, Al Qaeda was able to 'launch' effective attacks on urban targets in a country half a world away and, as noted in the Conclusion, it is likely that similar bands will use similarly inaccessible locations as the headquarters and even 'launching pad' for the apocalyptic attacks that will probably occur in the second quarter of the century.

However, the targets of these attacks and the operational zones in which insurgents deliver the attacks will be located in an *urban* environment. And the most obvious example of the influence of the physical environment upon insurgency is the difference between rural and urban insurgencies and between bandit-like and gangster-like operational formats. Urbanisation produces a man-made physical environment that in turn is accompanied by a distinctive human environment – living in large and dense groups. It has become an increasingly prevalent feature of the human condition as throughout the globe there are more and more city-dwellers, who share a distinctive physical and human environment that arises from this density of buildings and population. And the modern trend towards urbanisation, in terms of land area as well as population, provides prima facie evidence that wars of any kind are increasingly likely to be fought in an urban environment.[16]

Furthermore, while urban areas have proved disastrous environments for a maoist-model preparation strategy, they seem to offer a range of advantages to the third-phase insurgencies' pressure, propaganda and provocation strategies. According to geographers' definition, urban areas 'reduce the friction of distance' by drawing together buildings that house people, goods, services, machinery and information.[17] This high density of population and buildings offers insurgents the benefits of numerous potential supporters, extensive cover or concealment and an abundance of targets that are close to hand and often of 'practical and symbolic importance' to the rulers.[18] In addition to these political targeting and social support-mobilising opportunities, the urban environment provides some straightforward military opportunities:

> [Cities] provide a physical environment that favours insurgent operations by allowing them to capitalize on their advantages of flexibility, short-distance rapid mobility through different sorts of terrain, and the possibility of physical concealment (merging with the civilian population and the like), while minimizing their inherent disadvantages of lack of fire-power, inability to deploy large units, and lack of long distance mobility.[19]

A distinctive feature of physical concealment in urban areas is that it may take the form of hiding within or behind the crowds of a numerous and dense civilian population. Analyses of urban warfare repeatedly point out that the presence of so many non-combatants poses serious problems for troops fighting in urban areas.[20] And these non-combatants also provide tactical opportunities for insurgents seeking to carry out hide-and-hurt or hit-and-run attacks.

The cover and concealment provided by the urban physical terrain is similar to a combination of mountainous and jungle terrain. After driving Chechen insurgents out of the city of Grozny in 2000, the Russian army showed its dislike of these features by carrying out the urban equivalent of defoliation or deforestation: 'Any building large enough to hide a sniper was systematically dynamited'.[21] The urban terrain can also hamper the counterinsurgency operations of the state's forces. For example, 'narrow, irregular streets and spaces deny to the security forces a number of their inherent advantages – especially their capacity to manoeuvre and deploy disciplined units, and to bring their superior fire power to bear at crucial points'.[22] If the state's forces are engaging an insurgent band in combat, they will find not only that it is more difficult to deploy and manoeuvre large units than in open country but also that artillery and air attacks are less effective than in open country.[23] Furthermore, combat in urban terrain does not lend itself to the technological solutions that modern armies look to as an answer to military problems, and they do not have as great a technological edge or advantage over the insurgent as they do in rural environments.[24]

However, urban and rural environments are similar in one crucial aspect – diversity. 'Combat in cities is unique to the opponent, the geography, the campaign, and the political considerations of the conflict'.[25] Even the physical environment of cities varies more than might be expected in place and time. A famous example of the latter is the way that Paris and some other European cities were redesigned in the 19th century to incorporate long and wide boulevards. For it is often thought that this change was made to improve troops' access and fields of fire, and therefore removed 'the ruler's fear of the unstable mob that infested the medieval streets and alleys of the old town'.[26]

The geographical variation in urban physical environments includes some Third World cities that have particularly narrow, warren-like or maze-like streets. In fact the Algerian FLN attempted to use 'the narrow streets of the Kasbah in Algiers as a fortress' against the French colonial forces in 1957.[27] A very different sort of warren-like or maze-like urban area was exploited in 1963 by urban revolutionary insurgents in the Venezuelan capital of Caracas. It was characterised by the *ranchos* (camps) of shanty-town dwellings that extended up into the hills on the outskirts of the city and put intruding military units at a severe tactical disadvantage.[28] Other Latin American cities have developed similar shanty-town areas that are known by such terms as *favelas* (squatter towns) and *callampas* (mushrooms), and they can also be found in other parts of the Third World, such as the *gecekondu* (overnight creations) in Turkey.[29] By the 1990s counterinsurgency experts were warning that thanks to 'their warren-like alleys and unpaved roads' such shanty or slum areas had 'become as impregnable to the security forces as a rural insurgent's jungle or forest base'.[30] On the other hand, jungle or forest bases have seldom been impregnable to well-equipped, high-quality and numerous security forces, so it is not surprising that shanty slums and urban warrens have been neutralised by military countermeasures. In the Venezuelan case the *ranchos* were eventually controlled by the simple expedient of stationing large numbers of troops in these areas. In the Algerian case the Kasbah 'succumbed relatively easily' to attacks by high-quality security forces.[31]

What is more, the key variation in urban physical environments is that many Third World cities now have modern features that actually make them more accessible and controllable by counterinsurgent forces. In particular, modern urban motorways improve troops' access to the inner areas of the city to counterattack insurgent operations or to conduct search and destroy operations against insurgents.[32] Similarly, modern suburbs are serviced by an impressive road network, in addition to having houses that are often constructed from relatively flimsy materials and leave plenty of space between houses or clusters of houses.[33]

An increasing number of Third World cities will also benefit from some features of Western inner-city development that help to defend high-value targets against insurgents' car bombs or suicide attacks. A prominent example is the way that a defensive perimeter, the Ring of Steel, was established around the City financial district of London in response to the IRA attacks of 1992–3:

> Subsequently, strategies were devised for the City centered on the construction of roadblocks, the use of armed checkpoints, the development of a series of public and private CCTV [closed-circuit television] networks, a number of traffic restrictions, increasingly visible policing, and a series of private initiatives to increasingly fortify individual buildings.[34]

The 'fortification' of individual buildings in the inner-city area has been described by urban geographers and architects as the creation of 'citadels', but this development had pre-dated the terrorist threat and had originally taken the form of 'including more and more of the facilities necessary for daily life within the building itself. One will never have to leave the citadel for shopping, for meals, for entertainment, for personal services'.[35] In fact the World Trade Centre destroyed by the 9/11 attacks had pioneered this format, and its fate dramatically illustrated that some modifications would have to be made in the citadel concept to reduce the likelihood of terrorist attack. In addition to reducing their attractiveness as targets, new defensive measures are being introduced. These new urban citadels will not only be 'less ostentatious externally, less ultra-high-rise signature buildings' but also be 'more comprehensive fortified centers, with high-tech metal detectors, fingerprint card entry' and other 'barriers to easy access'.[36] But of course Western political leaders must still proclaim that terrorism will never succeed in making their citizens change their way of life.

The political and military environments

The political environment of an insurgency is in a sense more important than its physical environment because its political goals – and thus its nature as a revolutionary, separatist or expeller type of insurgency – are derived from its local political environment. To take an obvious example, only if they are being invaded, occupied or ruled by a foreign state will insurgents be fighting an expeller insurgency rather than a revolutionary or separatist insurgency. And this difference in the type of insurgency has huge implications for not only the opportunity variable but also the motive and means variables, especially the method-means' questions of strategy, format and mode that were discussed in Part I. In fact the most important aspect of Mao's environmental approach was his recognition of the differing nature and types of war, including the differing types of insurgent stateless warfare. His awareness of these crucial differences was evident both in his intellectual criticism of a mechanical approach to the problem of war and also in his practical awareness of the differences between an expeller insurgency against the Japanese invaders and a revolutionary insurgency against Chiang Kai-shek's regime.[37]

The other way in which the political environment has a fundamental effect on the insurgency is that there will not be any need for an insurgency if its political goals can be pursued peacefully and without resort to the 'other means' – organised violence. Only if there is some political reason for using organised violence to pursue their expeller, separatist or revolutionary political goals, will insurgents rise up against the foreign or local state. This was recognised by Guevara in his famous maxim, quoted in the introduction to this chapter, which pointed out that only if the peaceful means of pursuing revolutionary goals – 'the possibilities of peaceful struggle' – had been exhausted, would people have reason to resort to the organised violence of guerrilla warfare.

The same point can be made in terms of the political opportunity, or lack of opportunity, for an insurgency to be successful. As was noted earlier, Guevara's maxim implies that the presence of a democratic political environment reduces the opportunity for revolutionary insurgents to such a level that it will negate the other two variables, motive and means, in its insurgency calculus. Political scientists' research into the Latin American revolutionary insurgencies has confirmed that revolutionary insurgents have no hope of defeating a democracy and indeed seem to have no hope of defeating some varieties of *non*-democratic regimes. For example, Wickham-Crowley's research not only confirmed the Guevarist invulnerability of democracies but also discovered that some non-democratic regimes were much more vulnerable than others. While 'patrimonial praetorian regimes', like Batista's in Cuba and the Somoza family's in Nicaragua, are 'strikingly vulnerable to revolutionary overthrows', quite the opposite is true of 'collective military regimes allied to the upper class'.[38] This point was in turn confirmed by Goodwin's research but he described the vulnerable type of non-democratic regime as a 'neopatrimonial' dictatorship that has an infrastructurally weak state and is resorting to indiscriminate repression.[39] He also implicitly confirmed Guevara's point that the presence of even seriously *flawed* democracy can remove the political opportunity for a revolutionary insurgency to develop. In addition to agreeing that a democracy will not produce a major revolutionary insurgency, Goodwin pointed out that such an insurgency did not arise in 1980s Honduras at least partly because its 'very deeply flawed' two-party 'quasidemocracy' still allowed citizens to protest and win some concessions: 'Hondurans never felt the need to join or support revolutionaries in order to defend themselves or to improve their welfare'.[40]

However, Goodwin presented two cases that showed the limitations of *democratisation* as a means of removing the political opportunity for an *existing* revolutionary insurgency to persist. As was described in Chapter 4, his studies of the Salvadoran and Guatemalan cases showed that a revolutionary insurgency could persist even after a military-ruled state had apparently been democratised, because in these cases the security forces maintained a degree of indiscriminate repression that kept the insurgency alive until it could negotiate a compromise peace settlement with some credible security guarantees. Presumably the insurgency's political opportunity would have been more seriously weakened if the democratisation had included sufficient control of the security forces for the insurgents and their supporters to believe they could pursue their political goals through peaceful means. And presumably this analysis of leftist insurgents operating in Latin America can also be applied to Islamist revolutionary insurgents operating in the Middle East and Africa.[41]

But expert analyses or research relating to revolutionary insurgents certainly cannot be applied to separatist insurgents and their political opportunity in a democratic political environment. The obvious reason is the famous 1921 victory of the Irish separatist insurgency that arose and operated within Britain's democratic political environment. And there are several similar cases that have achieved the partial success of long-term persistence and/or a compromise peace settlement. Guevara's maxim helps to explain why a democratic political environment does not necessarily weaken the political opportunity of separatist insurgents to such a degree that it negates the other variables. His assumption that 'the guerrilla outbreak cannot be promoted' when 'the possibilities of peaceful struggle have not yet been exhausted' highlights the fact that separatists' possibilities of peaceful struggle have usually been exhausted quite early in the life of a democracy.[42] Dion pointed out in the mid-1990s that 'secession is difficult in well-established democracies', as a minority group wanting to secede will find that 'other citizens are generally hostile to their plan' and of course the minority cannot hope to win a majority in elections or a referendum.[43] She noted that in fact *no* democracy that has been established for more than a decade has ever experienced a secession by election or referendum.[44] So unless separatists are politicking early in the life of a democracy or will accept something less than secession, their possibilities of achieving their political goals through 'peaceful struggle' would appear to have been 'exhausted' before they even begin to struggle. What is more, separatist insurgencies seem unaware of the window of opportunity for peaceful struggle that exists in the early years of a democracy. For example, the Basque ETA continued its insurgency during the early years of Spain's post-Franco democracy, and the Kurdish PKK continued its insurgency during the early years of the new democracy established in Turkey after the 1980s military dictatorship.

As for expeller insurgencies, the 'possibilities of peaceful struggle' against a foreign state's rule can have a major effect upon the opportunity for anti-colonial insurgencies. It is true that the Western imperial powers, notably France and Britain, fought a number of counterinsurgency wars in the 20th century in an attempt to retain their colonial territories. But these wars were less common than the many instances of the imperial powers 'going peacefully', often after surprisingly little political pressure from their colonial subjects, and thereby deliberately or inadvertently removing the political opportunity for an expeller insurgency. The end of empire is perhaps better symbolised by Britain's peaceful withdrawal from India in 1948 than Portugal's willingness to keep fighting counterinsurgency wars in its three African colonies until the mid-1970s. The reasons for these differences in bellicosity are to be found in the imperial state's domestic politics as well as in the political environment of a particular colonial territory, and in this sense the political opportunity for an anti-colonial insurgency extends beyond the local environment to include the political environment of the foreign state that the insurgents are fighting against.

The same argument applies to other forms of expeller insurgency and in fact this extension of the political environment was implicitly recognised in Chapter 3 when discussing the pressure strategy that is so often used against foreign states. It was implied a) by Mack's highlighting of the asymmetrical difference between the insurgents' war of survival and the foreign state's limited war, b) by the issue of democracies' greater political vulnerability than non-democracies to pressure from colonial or other expeller insurgents and c) by the discussion of the political calculus of the pressure strategy – which is very much concerned with the insurgents' political opportunity to achieve a degree of success through the method of a pressure strategy. For example, in Chapter 7 the Somali warlord General Aidid was described as achieving success with a pressure strategy but this success depended on the political opportunity provided by the political calculus in US domestic politics. The withdrawal of US troops from Somalia in 1993–4 has been ascribed to 'the American Achilles heel – risk and casualty aversion'.[45] But it was also due to a political calculus that included President Clinton being able to wind up the Somali operation without loss of personal prestige because it was not him but President Bush who had been responsible for sending the troops to Somalia in 1992. If the political calculus had not been so favourable, the Somali expeller insurgency would have had a much weaker political opportunity that would have negated its pressure-strategy method. However, that lack of political opportunity would have probably been accompanied by a larger US military commitment that would have so weakened this expeller insurgency's military opportunity that its other variables would have been 'doubly' negated – by the weakness of its military as well as political opportunity.

Whatever the type of insurgency, it may well be foredoomed as much by a lack of military opportunity as a lack of political opportunity. When Goodwin was discussing the vulnerability of dictatorships to revolutionary insurgencies he included 'the requirement that a dictatorship be infrastructurally weak, for a powerful dictatorship with a reliable army could, in principle, repeatedly and effectively repress any emergent opposition'.[46] In fact a state is often able to repress an insurgency with its police forces rather than its military forces. As Galula pointed out, the police are the 'eye and arm of the government in all matters pertaining to internal order' and therefore are 'a key factor in the early stages of an insurgency; they are the first counterinsurgency organization' that has to be 'neutralized' by the insurgents.[47] Only if the police are unable to destroy or contain the insurgency, will the military be deployed in this matter pertaining to internal order rather than external defence. (Indeed a state may well prefer to contain an insurgency through its police force than to destroy the insurgency through its military forces and face the consequences of using the military in a constitutionally dubious and politically dangerous role.) Once the military is deployed against the insurgents, the calculus of insurgency is no longer required to make a hypothetical assessment of whether the latent strength of the military denies the insurgents any military opportunity of success. The calculus instead has to assess whether the military's counterinsurgency effort is in reality creating such an unfavourable military environment for the insurgency that its weak military opportunity will be negating the strength of its motive and means variables.

If the state's military forces are required to take on a counterinsurgency role, they normally have both the method and capability to create a very unfavourable military environment for the insurgents. The method is supplied by the impressive body of counterinsurgency 'doctrine' that has long been available. For example, the environmental approach was at least implicitly adopted and advocated as long ago as Callwell's 1896 classic *Small Wars*. He emphasised 'the extraordinary diversity of conditions which makes the consideration of small wars so complex', the 'great variety in the natures of enemy to be dealt with' and that 'in no two campaigns does the enemy fight in the same fashion' – in fact 'each small war presents new features'.[48] By the 1990s an explicitly environmental approach was literally the textbook approach. O'Neill's text on insurgency included under the headings of physical and human environment: terrain, climate, transportation-communications infrastructure, demographic distribution, social structure, economic factors, political culture and the political system.[49]

Furthermore, in the 1960s the environmental approach had been reinforced by a series of contributions that formed a formidable basis for counterinsurgency doctrine. Galula's book on counterinsurgency provided a counterpart of Guevara's handbook on guerrilla warfare and, for example, identified three distinctive features of counterinsurgency: 1) that there is an 'asymmetry' between the assets/liabilities of the insurgent and those of the counterinsurgent, 2) that the mind-sets and tactics appropriate for conventional warfare can be inappropriate for counterinsurgency and even be counterproductive and 3) that the population is the counterinsurgency objective, because without the cooperation of the population, it is impossible to prevent the insurgents from re-emerging (militarily returning and politically rebuilding) in an area from which they have been cleared.[50] A few years later

Draper pointed out the importance of the political factor, as a precondition of military victory over the insurgents, when he ascribed the French and American counterinsurgency failings in Vietnam to the failure to recognise that 'sound politics in Vietnam was the precondition of military victory, not that military victory was the precondition of sound politics'.[51] At about the same time, Thompson was pointing out the importance of what might be termed 'sound administration' in defeating communist insurgency. He emphasised 'the great advantage which the communists derive from weaknesses in the government', such as corruption, and declared that the 'correction of these weaknesses is as much a part of counterinsurgency as any military operation. In fact, it is far more important'.[52]

In addition to the availability of tried and true methods, the military is likely to have an impressive capability for carrying out these counterinsurgency methods. This capability can be described and assessed through the same PIE formula that was applied to insurgents' capability but in the counterinsurgency case the Information component has priority – the formula would be Information, Personnel and Equipment (IPE). The importance of information was highlighted by Huntington when developing the analogy between guerrilla warfare and submarine warfare that was described in Chapter 2. He pointed to the similarities between counterinsurgency and antisubmarine warfare, including the critical role that information plays in successful operations: 'The guerrillas and the submarines must be *found* before they can be destroyed. Sonar and radar are to anti-submarine warfare what spies and scouts are to counterguerrilla warfare'.[53] However, it is not so much spies and scouts as information provided by the local population that is the counterinsurgent equivalent of sonar. When a US military theorist on counterinsurgency served in Iraq in 2003–4, he discovered that his earlier 'ritual obeisance' to the importance of intelligence was 'truer than I knew at the time that I wrote it', but he also discovered that 'the task of winning and keeping the support of the population is far more complex than I had understood'.[54]

In fact this aspect of the Information component is related to the Personnel factor, for without sufficient quantity and quality of personnel it will be difficult to win and maintain the support of the population – or do anything else effectively in a counterinsurgency war. The quantity of troops required to defeat an insurgency can be dramatically large and is far more disproportionate than what is required to defeat a conventional military force. Galula suggested that 'unless the population can be trusted to defend itself' (such as through local militias), the military would need 'a ratio of force of ten or twenty to one between the counterinsurgent and the insurgent':

> The French forces in Indochina never approached this ratio, a fact that, more than any other, explains why the French could not have won there even if they had been led by Napoleon.[55]

As for the quality of the personnel, the disastrous effects of poor quality was famously illustrated by the failure of the sizeable Cuban army to deal with Castro's guerrillas, which it certainly outnumbered by Galula's ratio of ten or 20 to one. The Batista regime's army has in fact been described as 'less a fighting force than a uniformed extortion ring', which allowed the guerrillas to bribe or persuade the military's officers and units to desert or to surrender after only token opposition – fewer than 300 of the Cuban military died fighting to defend Batista's regime against Castro's guerrillas.[56]

The third component of the IPE counterinsurgency formula, Equipment, overlaps with the information factor when it comes to electronic information-gathering equipment – everything from satellite-surveillance to tracing mobile-phone locations. But only the US can deploy such an array, and for the average state it is such prosaic items of equipment as the helicopter that have added most to their counterinsurgency capability. For example, a major contributor to the Turkish military's defeat of the PKK's guerrilla war in the early 1990s was the increasing use of helicopters, particularly the introduction of night-capable Super Cobra attack helicopters.[57] And in the words of Porch's brilliant updating of the colonial-era jingle about the Maxim machinegun: 'Whatever happens, we have got/ The helicopter and they have not'.[58]

Finally, the lack of military opportunity includes the state's conventional, not counterinsurgency, military capability to defeat insurgents who seek to destroy it through semistate warfare. Clearly 21st century militaries should have no difficulty in destroying an insurgent regular army, and the only exceptions to this lack of military opportunity will appear in countries similar to Nepal. The military destruction of the Tamil Tigers' semistate in 2008–9 is indicative of the lack of military opportunity in most parts of the world for a maoist-model shift to semistate warfare.

Nonetheless, both conventional and counterinsurgency warfare have a long record of blunders and incompetence. Throughout modern history there have been instances of mistakes in method or shortcomings in capability that have given insurgents the military opportunity for victory or a substantial degree of success. And one of the indicators of whether the military are failing in their counterinsurgency effort is how much of the state's territory and population is to be found in O1 and O2 rather than O3 and O4 operational zones.

Operational zones

The operational zones form a layer of 'very local' military-political environments, within the insurgency's overall military environment, which can vary markedly in type and in the amounts of territory or population that they cover. The variation between types of zone and the implications for insurgents' methods and operations is illustrated by contrasting Kalyvas's description of a zone where the state's forces are in full control with his description of a zone where the insurgents exercise a control close to that achieved in a semistate base area. In the latter, the insurgents dominate the area but 'cannot prevent sporadic visits by incumbent forces' who may well leave within a few days; in a zone where the state's forces are in full control, the insurgents have few or no clandestine cells to help them, will 'lack access to people and information' and will find travel 'difficult and fraught with dangers' – they will be capable only of terrorist operations and acts of indiscriminate violence.[59] These are only two of the various types of operational zone that may be found within a particular insurgency's military environment. In many cases of small-scale and terrorist-mode insurgency virtually the whole country will be under the state's full control. But in large-scale insurgencies there are marked regional and/or rural-urban differences and the variation can become so local that there are significant differences from one village to the next or one neighbourhood to the next. From the counterinsurgency perspective Galula described this situation as 'a patchwork of pieces with one well pacified, next to another one not so pacified or perhaps even under the effective insurgent's control'.[60]

The typology used in this chapter and the next to describe the various types of zone is based upon Kalyvas's five-zone analysis of the logic of violence in civil wars. His civil-war perspective on insurgency means that he is focused on the maoist-model preparation strategy and on Trotsky's depiction of a revolutionary situation as a dual power/sovereignty that, as was described in Chapter 3, eventually takes a territorial form as a civil war between semistates.[61] Kalyvas takes the notion of contestation of sovereignty a step further:

> The simplest way to conceptualize the division of sovereignty in civil war is to distinguish between zones of incumbent control,

zones of insurgent control, and zones in which control is contested.[62]

But instead of this simple three-zone typology he provides a more nuanced, *five*-zone conception of the distribution of control between insurgents and incumbents as it varies across the continuum of zones. The continuum extends from the incumbents' uncontested, 'full' control in zone 1 to the incumbents' incomplete but secure (dominant) control in zone 2, then the intermediate, more equal or truly 'dual' control in zone 3, the insurgents' incomplete but secure (dominant) control in zone 4 and finally the insurgents' uncontested, 'full' control in zone 5 – their base areas or liberated zones.[63]

In addition to their political dimension of sovereignty/control, these zones also have a crucial military aspect that is highlighted by the way in which they can quickly change from one type to another. The shifts in degree of control and type of zone 'are primarily a function of tactical military decisions' regarding 'how to allocate scarce military resources', such as carrying out a clear-and-hold counterinsurgency campaign to establish uncontested control over an area that had previously been under dominant or equal control.[64] The five zones identified by Kalyvas are therefore actually five different types of military-political environment that occur in practice as different very-local environments within a local military environment.

Four of these five zones will be relabelled and renumbered as insurgent *operational* zones. Zone 5 is excluded because insurgents' base areas or liberated zones are not normally one of the scenes of their military operations, but zone 1 is included because some forms of insurgent operation are sometimes carried out or attempted in areas controlled by the state's army, administration and police. The four operational zones are described in Table 8.1, which is derived from Kalyvas's control-based definitions and his descriptions of what these levels of control mean in practical, operational terms. But his numbering of the zones has been reversed, so his zones 4, 3, 2 and 1 become respectively O1, O2, O3 and O4.

In some insurgencies, control has been exercised by neither insurgents nor the state but by such 'third parties' as tribal leaders, warlords, local notables or even bandits who are allies of the state or are virtually independent 'local strongmen' – and indeed there may be anarchical areas of a country in which not even third parties exercise control. Furthermore, as was noted earlier, the police counter and often defeat or contain an insurgency in its earliest stages and, whether in these

Table 8.1 Insurgents' Four Operational Zones: O1 to O4

	Control	*Insurgents' Presence*	*State's Presence*
O1	dominant	semi-liberated zone	sporadic military visits
O2	equal/dual	move freely; night-time control	continuous, especially daytime
O3	little	within reach; clandestine cells (and sporadic night-time visits)	dominant control
O4	none	few or no clandestine cells	uncontested, 'full' control

stages or after the military have regained control of an area, the state will tend to rely on its civilian police and intelligence services rather than military forces to maintain control in O4 and perhaps even O3 zones. Somewhat paradoxically therefore an indication that an insurgency is being defeated or contained is that it does not have any truly 'military' operational zones.

The particular geographical distribution of types of zone within an insurgency's military environment is one of the distinctive features of its local military environment. And obviously an insurgent band will adjust its strategy and other methods to fit the particular types of military or political opportunity that are offered by particular types of operational zone. For example, establishing a base area or training recruits might be appropriate in an O1 and perhaps an O2 zone but is much less opportune in an O3 and simply suicidal in an O4 zone. On the other hand, not only propaganda but also provocation strategies might well be opportune in an O4 zone and might also involve changing to a terrorist mode of operation.

Furthermore, there is another form of military opportunity that varies markedly according to the type of operational zone, namely the opportunity to elicit or extract *social support* from the local population. Mao's emphasis on guerrillas' need for social support, as encapsulated in his legendary fish-in-the-water analogy, is of course applicable to not just maoist-model revolutionary guerrillas but all insurgencies and insurgent strategies. Yet even Mao's mobilisation of support from the peasantry 'was considerably more coercive than was previously thought' and there is a need 'to decrease the roles ascribed to popular participation or spontaneous action' in winning that support – instead 'the importance of local military superiority is almost self-evident'.[65] This local military superiority is characteristic of insurgents' semistate base areas and O1 operational zones, in which they can use coercion to help mobilise social support. The notion that such support can be motivated partly or wholly by coercion may seem paradoxical but 'support' for an insurgency should be objectively defined in terms of the actual *behaviour*, not subjective attitude or feelings, of the members of society who 'deliver the goods' to the insurgents. Of course the insurgents will have some genuine political supporters who favour the insurgents' political goals and, as was described in Chapter 6, have a selective incentive to contribute to this collective good. But some of the local population who are subject to insurgents' sovereignty/control are what will be termed 'fearful subjects', who are coerced into providing support even though they feel neutral or hostile towards the insurgency.

In addition, there may be some 'local allies', who feel neutral or hostile towards political goals of the insurgency but provide it with support quite voluntarily because they share a common enemy and/or are receiving benefits from the insurgents. Kalyvas's recent explanation of such local-level alliances depicts them as hard-headed exchanges that are based upon insurgents' military contribution to their allies' local conflicts. He gives many examples of local cleavages and conflicts – between towns, villages and especially families or kinship groups – that led to one side or other of the local conflict becoming involved in an alliance with supra-local rebels with a revolutionary, separatist or expeller cause.[66]

> Alliance entails a transaction between supralocal and local actors, whereby the former supply the latter with *external muscle*, thus allowing them to win a decisive advantage over local rivals; *in exchange*, supralocal actors recruit and motivate supporters at the local level.[67]

But, like the fearful subjects, the local allies will not be interested in supplying support unless the insurgents have the local military superiority that enables them to coerce fearful subjects and to lend some external muscle to local allies.

Consequently, wherever the insurgents have the military superiority associated with a semistate base area or O1 operational zone, they also have the opportunity to extend their base of support by extracting it from fearful subjects and eliciting it from local allies. In contrast, within an O4 operational zone – which is under the state's uncontested control – the insurgents have to rely solely on the more genuine support delivered by political supporters. Furthermore, it becomes increasingly difficult to elicit, give and receive support as the insurgents' level of control decreases and the state's increases, so the actual mechanics of delivering support become increasingly problematic in O2, O3 and O4 zones. Clearly, the distribution of types of operational zone, such as the extent and location of O1 and O4 zones, within the insurgents' local military environment will have a marked effect upon the amount of support that is delivered to the insurgents.

More generally, the nature and level of social support delivered to insurgents is a result of the interaction between the military environment and the social-political environment. As the local social-political environment may have just as many very-local variations as the military environment, the interaction between them may well produce some complex or eccentric very-local results, and it is hardly surprising that the support for insurgents can vary from one hamlet to the next in rural areas and from one street to the next in urban areas. Yet an insurgency's social support is best described as its support 'base' both because it is the foundation of the insurgency's military capability and because this social support plays as important a role in stateless warfare as supply bases do in warfare between states. If it is true that 'amateurs talk tactics but professionals talk logistics', this is also true in stateless warfare. However, the professionals will be talking about the nature of the social support base in a particular operational zone or in the insurgency's overall local environment, for the nature of the support base will affect not only the insurgents' tactics but also their operations, strategy and perhaps even format or mode. So although the social support base is the focus of Chapter 9's analysis of the social-political environment and opportunity, it is also part of this chapter's analysis of the military opportunity offered to insurgents by their social-political environment as well as their military-political environment.

Support bases

The social support base for an insurgency can be visualised as a foundation that has three different aspects – depth, breadth and strength. The more active the form of social support, the 'deeper' the support base. The greater the proportion of a society that supports the insurgents, the 'broader' the support base. And the more the support is motivated by political support for the insurgency, the 'stronger' the support base.

The depth of a social support base involves two different levels of support that O'Neill described as passive and active.[68] Passive support is merely the denying of information to the insurgents' enemy, namely a state's counterinsurgent forces. Information is a crucial aspect of counterinsurgent capability, as was emphasised earlier in the chapter, and therefore the *denial* of this information greatly increases the insurgents' military opportunity. In fact Begin suggested that the denial of information – 'silence' – is probably the most important aid that the support base can provide:

> They helped the rebels. They saw them going out to fight and kept their lips sealed. They saw them coming back from battle and were silent. *Silence* was probably the most important, though not the only aid they gave. The people gave the underground what the country's natural conditions failed to give: *cover*. We did not hide behind trees; we were guarded by living trees. Otherwise we could not have fought, certainly could not have won.[69]

But such passive support is quite different from active support: 'intelligence information, concealment [in their homes], shelter, hiding places for arms and equipment, medical assistance, guides, and liaison agents'.[70] In this case the information is being provided to the insurgents rather than merely denied to their enemies and, as was described in Chapter 7, such information is an important factor in insurgents' military capability. The material rather than informational aspects of active support make the less direct contribution of providing a logistical foundation for the insurgents' capability.

This logistical rather than informational support includes everything from food or shelter to actual financial contributions. For example, the communist insurgents in 1950s Malaya were so dependent upon food supplies from supporters among the Chinese rice-growing population that one of the most effective counterinsurgency strategies was imposing a food-control system on the civilian population and conducting food-denial military operations.[71] In more recent times logistical support for insurgents has often been more financial than physical and has included such 'globalised' versions as the LTTE Tamil Tigers' attempt to institute a 'tax' on members of Tamil diaspora or migrant communities, with those living in affluent Western countries being individually liable for an annual payment of about 200 pounds sterling.[72] Perhaps the best-known form of financial support is the financing of Al Qaeda by donations from sympathetic individuals, mosques and charities but in some cases this was possibly inadvertent or under false pretences and therefore is difficult to categorise.[73]

The insurgents' support base also provides such human assistance as the 'guides and liaison agents' mentioned earlier and, most importantly, a recruiting base from which new recruits come forward to join the band – for motives that were discussed in Chapter 6. However, the boundary between this part of the support base and actual membership of a band is often blurred, particularly when the insurgent band includes part-timers. For example, many of the jihadists in Sageman's sample were employed in normal careers or marginal jobs, were some kind of student or were unemployed and living off welfare benefits or petty crime.[74] Another example is the Basque ETA's different levels of membership, with the great majority of members being only part-timers who might take part in only one or two attacks a year and perhaps in a support role rather as an armed participant.[75] A different sort of blurring of the line between member and supporter is evident in Petersen's description of the second level or node of roles in a rebellion, which was mentioned in Chapter 6. This level or node involves either providing direct support to the rebels or participating in a part-time locally based insurgent band – as distinct from full-time membership of a mobile guerrilla unit or the rebel army.[76] The distinction is similar to that between a) full-time mobile columns and b) part-time local guerrillas in the contemporary Taliban insurgency in Afghanistan, and shows that either an insurgent band has to be viewed as having two levels of membership or that the support base has to be viewed as containing a level of *very* active supporters who are willing and able to join in the fighting:

> Full-time fighters (sometimes called Tier 1 Taliban) are usually therefore foreigners in any valley where they operate, unless it happens to be their actual home valley, and a mobile column is usually made up of people from several tribes, reinforced by foreign fighters ... Local guerrillas (also known as Tier 2 Taliban) ... fight almost entirely in their home valley, usually within a day's march of their village, close to the roads, on a relatively casual 'pick-up team' basis, and almost always in support of the full-time column.[77]

However, depth needs to be combined with sufficient breadth, with a broad enough section of society or of a large enough social group. As Begin implied, a broad but relatively shallow support base may be more effective than a deep but narrow support base, with active support being provided by only a small minority and the rest of society providing not even passive support. An example of the vulnerability of the overly narrow support base is the demise of revolutionary bands and microbands that began operating in the West in the 1960s–70s, such as the Baader-Meinhof group in Germany or the Weathermen in the US.

They were operating in an almost entirely O4-zone environment and were given active or passive support by only a miniscule section of society. As Gurr pointed out, 'group support' was 'essential to sustain terrorist activities' in a democracy, because without it any urban band 'finds it increasingly difficult to attract new recruits, to get material resources, to find refuge among reliable sympathizers, or to avoid informants'.[78] And if the support base is too narrow, the terrorists are vulnerable not only to being killed or captured, as was the Baader-Meinhof group, but also to 'fading away', as did the Weathermen, thanks to a lack of recruits and a lack of motive-to-continue among the remaining, over-stressed members. The Weathermen band 'never recruited more than four hundred members and followers, and most of its inexperienced leaders and recruits were worried not about the revolution but about their hideouts, survival logistics and internal group dynamics'.[79] After their final public manifesto was published in 1974, nothing significant was heard from them and so it seems that the surviving members of the band opted for what they presumably believed was a hard-earned retirement from insurgency.[80]

Table 8.2 The Social Support Base

Depth		Breadth	Strength
Passive Support Denying information to enemy = 'silence'	*Active* Support 1) information 2) logistical 3) recruits	Proportion of population	1) fearful subjects 2) local allies 3) political supporters (+ selective incentives)

Finally, the strength rather than depth or breadth of the support base will be increased the more it is composed of political supporters rather than local allies or fearful subjects of the insurgents. Political supporters are similar to politically motivated insurgents but lack sufficient motive to join the band or simply lack the means or opportunity of joining, such as being too old or young to be a combatant. Their strength of commitment reduces the brittleness of a support base, and is why the narrow support bases in O4 zones are much more resilient than the wider support bases in semistate or O1 zones, which tend to disintegrate if the 'fortunes of war' lead to the insurgents losing their local military superiority. The lack of strength in the support base also creates the sorts of problems faced by the communist insurgents in late 1940s Greece. Their lack of political support meant that they were having to force villagers to join guerrilla units and were hampered by informant and supply problems: 'their infiltrating units had to hide from the local population when they could not cow it, and their operations lasted generally as long as the supplies they carried with them'.[81] A different sort of problem was faced by UNITA forces in late 1990s Angola, who carelessly lost much of their political support and were unable to regain it after the 'fortunes of war' turned against them. They had won the political support of the local population when dependent upon them for food and other needs but after the conquest of diamond-mining areas, the insurgents maltreated the population because 'those commodities could now be purchased abroad with diamonds'.[82] However, UNITA lost its hold on the diamond-mining areas after a disastrous shift to semistate warfare and was faced with the problem that 'given the rebels' violent track record, the rural populations were particularly loath to aid them, making it impossible to return to the classical Maoist framework of guerrilla warfare'.[83]

9
The Social-Political Environment

In this chapter the focus is still on the social support base but from the perspective of social rather than military opportunity – and of the social-political rather than military-political environment. As was described in Chapter 8, the passive or active support that is provided by fearful subjects or local allies is largely related to the military environment, especially the fact that these people are living in the insurgents' semistate area or an O1, O2 or even O3 operational zone. In contrast, the passive or active support provided by political supporters is largely related to the social-political environment and the social opportunity it offers the insurgents to convert *potential* political support into actual support. To exploit this opportunity the insurgents have to win support for the insurgency's political goals as well as use coercive, social or moral selective incentives to overcome free-rider and other collective-action problems, as was described in Chapter 6. If the insurgents are willing and able to exploit the opportunity, they can expand/deepen/strengthen their support base through an increase in the number of people supplying support, through transforming passive into active support or through converting fearful-subject or local-ally forms of support into the more resilient and reliable form that is provided only by political supporters.

This chapter therefore discusses the key issue involved in assessing the social opportunity aspect of the overall opportunity variable – how much potential political support was or might be offered by the social-political environment. From the subjective perspective of the participants there are also the issues of how to exploit this social opportunity and how much time, effort and risk should be invested in exploiting the opportunity to convert potential into actual support. From the academic, objective perspective the issue is no more than a matter of using sociological theory and research to explain or predict how much potential support was or would be available to the insurgents. But this is a very difficult problem, considering the diversity – in time and place – of the local environments in which insurgencies have occurred or might occur in the future. Of course part of the diversity is due to the variations in the military environment but there is a range of other possible influences on a local environment. In his analysis of 1960s Latin American insurgencies, Gott highlighted the diversity of their local and very-local environments: 'The more one looks at the problem, the more one is led to conclude that, in fact, conditions vary not just from country to country, or from province to province, but from valley to valley. Guerrillas may find support in one hamlet but not in another'.[1] It is not surprising that historians are now studying what they describe as the *social ecology* of revolution, which 'is concerned to relate local conditions to wider processes of social and political change'.[2] They are showing that Mao's insurgency, too, had to deal with a marked variety of local environments even within the communists' base areas in northern and central China. Studies of the different base areas are revealing 'immense variety and geographical specificity', and even within the main Yenan base area 'there was considerable variety, ecological and sociopolitical'.[3] Similarly, political scientists studying African guerrillas have recently endorsed political anthropologists' scepticism about the academic goal of 'generalisability' and have espoused an approach that is 'historically grounded and integrates multiple levels of analysis, from the local and national to the regional and global'.[4]

Nonetheless, an attempt to explain or even predict the actual and potential political support for insurgency is the best opportunity for sociological theories and research to be applied to the study of insurgency. These theories and research are much more applicable to large sections of society than to insurgent bands that number only hundreds or merely dozens of people. Laqueur pointed out in his first book on terrorism that social scientists 'may (or may not) explain the rise and decline of mass movements, but they cannot possibly predict the behaviour of a handful of people'.[5] As he argued, social and economic trends cannot explain the behaviour of members of very small groups, which are much more affected by such accidental factors as the presence of an outstanding leader than are large-scale political movements.[6] Similarly, Sageman's study of jihadists' motivation found that the social-science theory of relative deprivation was useful but ran into 'the fundamental problem of specificity' – many people suffer from relative deprivation but 'very few of them become terrorists'.[7] The problem of specificity means that any general, sociological explanations of why some people adopt insurgents' ideologies or political goals cannot explain why a few of these individuals go on to become insurgents. The best that can be hoped for is an explanation 'in the statistical sense that at least a very small random number of individuals might be motivated to carry this ideological commitment to its extreme'.[8]

But better things can be expected of explanations of why hundreds of thousands, perhaps millions, of people have become passive or active political *supporters* of an insurgency. For example, the theory of relative deprivation that Sageman found of only limited use when applied to insurgents is likely to be much more useful when applied to a large support base. It might reasonably be expected that a sociological theory could explain why the insurgency had the social opportunity to develop such an impressive support base – why the social-political environment provided the insurgents with so much *potential* support to convert into actual support.

Rebels with a cause

The obvious place therefore to begin is with Gurr's classic relative-deprivation theory, presented in his 1970 book *Why Men Rebel*. Although there is an 'infinite variety of motives' for rebellion, the key motivational factor arises when people perceive a 'discrepancy between the goods and conditions of life they believe are their due, and the goods and conditions they can in fact get and keep'.[9] He emphasised that this *relative* deprivation is related to aspirations that are perceived to be realistic and attainable. The discontent is 'not a function of the discrepancy between what men want and what they have, but between what they want and what they believe they are capable of attaining'; the nature of aspirations also helps to explain why the same concessions can mollify people who are fighting to preserve what they have but will be insufficient to mollify those fighting to satisfy new or intensified aspirations.[10]

Furthermore, these aspirations may be focused on something other than just economic or material conditions of life. For people also have aspirations for 'security, status, a sense of community, and the right to manage their own affairs', and in fact once 'basic physical needs are met, provision for these aspirations is at least as important as increased material well-being'.[11] Thus Gurr's theory can be applied to not just revolutionary but also separatist and expeller insurgencies, where the aspirations are focused on a nation's or ethnic/sectarian group's 'right to manage their own affairs' and to attain other non-material 'goods'.

The perception of deprivation relative to what is due to them may well lead to the deprived becoming the rebellious. Such relative deprivation produces a discontent that may be sufficiently intense to produce a motivational 'disposition to aggressive action', which is directed towards political actors if they are perceived to be responsible for the deprivation.[12] So the social-political environment is offering insurgents a major opportunity – sufficient potential support to build a strong as well as broad support base. But to exploit this opportunity the insurgents will have to complete a two-step process. First, they will have to ensure that the discontent becomes a *political* grievance by being directed against political actors, specifically the local or foreign state that the insurgents are fighting against. Secondly, the insurgents will have to ensure that this political grievance is transformed into *support for the insurgency's political goals* because they are the best way of attaining a collective good that will satisfy this political grievance.

Such a process of converting potential into actual political support is similar to what Galula described as converting a *latent* opposition into an acute and valid cause for rebellion. He emphasised that insurgents needed to be 'rebels with a cause' – that an attractive cause was a prerequisite for a successful insurgency – and he presented an almost sociological analysis of 1940s Greece to confirm that its communist insurgents were unable to win popular support because they lacked a valid 'cause' to espouse and champion.[13] What is more, he argued that the social problem or other type of problem that was the basis for a valid cause might be merely 'latent' and have to be made acute by 'raising the political consciousness' of that social group and its 'latent opposition'.[14]

This conversion of latent opposition into acute cause is similar to converting potential into actual political support but seems to involve only the first of the politicising steps described earlier – from relative deprivation to political grievance. It must be combined with the second step, from political grievance to support for the insurgency's political goals, before the process can produce political support for the insurgents. For example, both steps are evident in Weinstein's description of Sendero Luminoso (Shining Path) insurgents selling their rebel cause and goals to indigenous peasants in the Peruvian highlands.

> Sendero arrived making promises about how it could improve the lives of the often-ignored Quechua peasants. At people's assemblies, the rebels railed against a government that had left the highlanders

> without access to health care, education, transportation or sufficient food to survive in the rough climate. Peasants initially welcomed the Senderistas and their promises about the future as a possible avenue of escape from the struggles of their daily lives. ... [For] the *cause* was perceived as just and the *goals* reflected the desires of many peasants.[15]

Another way of describing these two aspects of politicisation is in marketing or sales terms. There is the demand-side creation of a demand for a political grievance to be dealt with and then the supply-side presentation to the customer of political goals and agents that will meet the demand. When converting potential into actual political support, the insurgents seem to have much to gain from carrying out their own demand-side politicisation because it gives them the chance to shape the new political grievance/demand to fit the insurgency's supply of political goals and agents.

Table 9.1 Converting Potential into Actual Political Support

1) Potential Support Relative Deprivation leading to Discontent and perhaps Political Grievance (Rebel 'Cause')
2) Politicisation: Demand-side and/or Supply-side Demand-side = Discontent into Political Grievance (Rebel 'Cause') Supply-side = Political Grievance into Political Support for Political Goals of Insurgency
3) Political Support + Selective Incentives = Support Delivered

However, demand-side politicisation may require much time and effort, even to the extent of dealing with that aspect of individuals' relative deprivation and discontent that Gurr described as the discrepancy between 'what they want and what they believe they are capable of attaining'. This may well require the insurgents to lead or even organise the local population into achieving local-level collective goods and goals to show these individuals what they are capable of attaining through contributing to collective action. (And such participation in collective action is also likely to strengthen the social selective incentives that the local community can use when it delivers support to the insurgents.) This is one of the reasons why the peasantry's contributions to the communist revolutions in China and Vietnam were based, as Taylor argues, on the communists' success in not only breaking the hold of local notables but also working with the peasants to bring about local collective goods, such as irrigation improvements.[16]

Furthermore, the insurgents may have to adjust their 'product' to overcome 'customer resistance'. Popkin's account of communist attempts to organise peasant villages in French-ruled Vietnam shows the communists stripping power from local notables and organising the peasantry for local collective action but also having to enhance their credibility with the peasantry by adjusting the communist party's vision of the future, which shifted from social revolution to 'cultural themes and nationalist appeals' that were centred on expelling the French.[17] However, if the product is adjusted too much to fit the local mindset, there is the danger of the insurgency's political goals becoming too localist and suffering from the Zapata syndrome. During the 1913–15 period of the Mexican Revolution the Zapata-led peasant movement in the Morelos region of Mexico was both a very strong and very weak revolutionary force. The movement's regional goals of regaining village communal lands expropriated by the hacienda class – and expropriating hacienda lands for redistribution among village communities – were supported by 'strong' village communities that could use social selective incentives to ensure individual villagers contributed to achieving this collective good.[18] The Zapata movement and its guerrilla-based army was therefore able to achieve its regional goals and control the countryside but was also not much of a revolutionary force outside its region, as the peasants were not very interested in what was happening in other parts of the country and the peasant-recruited army was unwilling to leave Morelos for any extended time.[19]

To avoid the Zapata syndrome any revolutionary insurgency has to extend its goals beyond regional or local interests to include a wider, national-level political goal. So although academic experts on revolution have pointed out that 'revolutionary movements are much more than simply ideological movements', these movements must have some ideological or national-level political goals if they are to avoid the Zapata syndrome.[20] It is true that guerrilla movements not only have to offer 'selective incentives to encourage participation' but also have to focus on local and immediate collective goods, such as law and order or land reform, 'before attempting to mobilize the population for more difficult tasks – including, ultimately, the overthrow of the incumbent regime'.[21] Yet if the focus on local goals becomes too strong, it may lead to the tail wagging the dog and to the incumbent regime being protected by the Zapata syndrome. Communist insurgents seem to have been well aware of this danger, presumably because of their strong ideological orientation, and even insurgencies that seem ideologically primitive when compared to the communists may be able to put together an effective package of national-level political goals. For example, the Taliban apparently adopted in 2006 a series of slogans as 'a rudimentary political platform for insurgent activity':

> These were 'Our party, the Taliban'; 'Our people and nation, the Pashtun'; 'Our economy, the poppy'; 'Our constitution, the Shari'a'; and 'Our form of government, the emirate'. ... The key strength of this approach is that it provides a series of unifying 'rallying calls' that appeal to a broad base of popular support by speaking to a wide range of aspirations and discontents – Pashtun nationalism, religious conservatism, the desire of narcotics traffickers to be left alone to pursue their activities, impatience with foreign occupation.[22]

The slogan that refers to the Pashtun as being our people and nation seems to confirm that the Taliban are becoming more of a Pashtun-separatist than Islamist-revolutionary insurgency but even separatist movements often have to be wary of localist tendencies, especially when these are based on strong tribal/clan loyalties. In these circumstances the main problem plaguing demand-side politicisation is not that of supplying local collective goods (and certainly the tribes or clans can deploy social selective incentives) but rather of overcoming localism and creating an overriding ethnic/sectarian grievance and separatist demand.

Yet such demand-side problems may not be any greater than those posed by a solely supply-side politicisation. For there can be serious problems involved in having to 'sell' the insurgency's political goals as the best way of dealing with an *existing* political grievance, which is already acute rather than latent. One of the most serious problems is having to deal with market competition a) from other bands selling other political goals as the way to satisfy the political grievance and b) from political parties selling other, non-violent ways of achieving political goals. In some circumstances an insurgent band need only raise the banner of revolt in order to attract political supporters but will find that *other* bands and organisations are raising *their* banners, too, and creating some confusion among the aggrieved about which banner to follow. Expeller insurgents have often faced such competition, as in the case of the different Palestinian bands and organisations or of the different Afghan mujaheddin bands that attempted to expel the Soviet presence. Competition has also appeared at times in separatist insurgencies, as in Northern Ireland between the Official and Provisional IRA, and in revolutionary insurgencies or movements, such as the several different Marxist bands operating in Brazil in the late 1960s or the many different Marxist and Peronist bands and parties seeking to end military rule in early 1970s Argentina.

Nonetheless, there have been occasions when an insurgent band has both enjoyed a monopoly of rebellion and been able to sell its rebel cause to virtually all potential consumers. Galula acknowledged that there are historical instances of a virtually unanimous majority of the population being 'deeply and genuinely attached to the insurgent's cause'.[23] For example, in 1950s Cyprus the Greek-Cypriot population seems to have unanimously supported the cause of the EOKA insurgents – to expel the British and become citizens of Greece. A British counterinsurgency analysis recognised this support for the insurgents' cause and noted that the British were therefore unable to employ successfully in Cyprus the practices that they had developed and used so successfully in Malaya, namely the 'hearts and minds' approach, the techniques for encouraging anonymous denunciations and the use of financial rewards for information.[24] In such cases as Cyprus the insurgent band seems to have been wholly successful in exploiting a huge social-political opportunity to convert potential into actual political support.

But building a deep as well as very broad support base requires the band to use selective incentives to overcome the same sort of free-rider and other collective-action problems that are involved in recruiting individuals to join the band. Although the rebel 'cause' – the political grievance and goals – gives the band's political supporters a collective motive, there are still collective-action problems to be overcome through the use of selective incentives, such as coercion, social pressures, a moral code's obligations or even affective relationships amongst close friends. Like the members of the insurgent band, the insurgency's political supporters are motivated not only by the collective motive but also by individually applicable selective incentives. Those used by insurgents to secure support are likely to take the form of intimidatory coercion, social pressures or some sort of moral incentive. In particular, coercion has the advantage of being both an effective form of selective incentive and an effective means of extracting or guaranteeing support from individuals with whom the band is *not* popular – from individuals who are neutral or opposed to the band – but are subject to its coercive rule and comprise the 'fearful subjects' component of the insurgents' support base.

For obviously it is more the exception than the rule to find insurgents enjoying the virtually unanimous political support of the local population. In fact the standard counterinsurgency assumption is that only a minority of the local population will favour the insurgents, the majority will have a neutral attitude, and another minority will favour the counterinsurgent side.[25] Similarly, the present conventional wisdom among academic experts, as summed up by Kalyvas, assumes that in a civil war or insurgency only a small minority of the population have a 'deep and unflinching commitment' to either side; most people 'appear to display a combination of weak preferences and opportunism' that might well be categorised as a 'neutral' attitude.[26]

The insurgents are best able to use coercion in a semistate area, where they exercise uncontested control over the local population. In this situation they can ensure compliance with their rule through state-like coercion that is based upon a selective use of violence.[27] Of course this coercion is claimed to be legitimate, for the insurgents not only are engaged in such 'ruling' activities as keeping order and extracting taxes but also are claiming to be the state-like legitimate rulers of the people. Yet whether or not the claim to legitimacy is accepted, the state-like coercion provides a way of motivating fearful subjects to deliver support as well as providing a coercive selective incentive (in addition to moral or social selective incentives) that deters political supporters from free-riding on one another's contributions.

But fearful subjects are less likely than political supporters to supply useful information or useful recruits and they are also less productive suppliers of logistical support. Fearful subjects coerced into providing logistical support are likely to fight back with what Scott described as 'the ordinary weapons of relatively powerless groups: foot dragging, dissimulation, false compliance, pilfering, feigned ignorance, slander, arson, sabotage and so forth'.[28] His analysis of how these 'weapons of the weak' are used as everyday forms of peasant resistance can also be applied to other relatively powerless groups, and the effectiveness of these 'weapons' is highlighted by the fact that even dictatorial states prefer to persuade rather than coerce their subjects.[29] Clearly, a much more effective and efficient form of active support is delivered by individuals who are persuaded political supporters, not fearful subjects, and who are being coerced only to the limited extent required to overcome collective-action problems. In fact political supporters are likely to engage in a form of collective self-motivation that has been described as 'peer monitoring', in which a peer group or a community bestows social rewards or punishments on fellow members who are known to be, respectively, conscientious or shirking.[30] So it is not surprising that apparently the mixture of coercion and *persuasion* is particularly prominent in base areas or liberated zones, and that some inhabitants of these semistate areas may actually 'switch their preferences' to favour the insurgency because they are benefiting from the band's attempt to win their political support through such measures as rent reductions or land reform.[31] These measures may also include local political reforms, as the SL in Peru and NRA in Uganda apparently secured the political support of local communities under their control by instituting a degree of participatory and power-sharing democracy at the grassroots level.[32] Such investments of time, effort and risk in converting potential into actual political supporters are likely to prove very profitable if this allows the insurgents to switch more of their military resources to fighting the enemy rather than intimidating the local population.

However, a very different situation and assessment occurs when the insurgency is located not in a semistate area but in operational zones. Conversion of potential into actual political support through demand-side or supply-side politicisation becomes increasingly difficult as the type of operational zone changes from O1 to O2 and then to the highly dangerous O3/O4 zones in which the state has dominant or full control. And the third-phase insurgencies normally will be located in operational zones rather than semistate areas, which are characteristic of the outdated maoist-model rural insurgency preparing for a shift to semistate warfare. The third-phase insurgencies are pursuing a pressure, propaganda or provocation strategy, which does not require a semistate area, and they are often operating within an urban environment in which a semistate area is a suicidal anomaly.

These insurgencies are likely to use an O1 zone as the less vulnerable equivalent of a semistate area. The O1 zone has most of the advantages of a semistate but allows the insurgents to 'cut and run' when faced with overwhelming force – and return when the state's soldiers return to their bases or move on to their next area of operations. Such temporary withdrawals by insurgents occurred in some urban areas of Iraq in 2003–6, and a travelling run-and-return system has been adopted by the Taliban in Afghanistan. As was mentioned in Chapter 8, the contemporary Taliban has full-time insurgents operating as flying columns and supported by part-time guerrillas based in their local valley and village; in combination they form what Kilcullen terms a 'Taliban operating system' that seems to provide an O1 level of control and an opportunity for demand-side or supply-side politicisation. The flying column engages mainly in 'political and religious indoctrination, armed propaganda, intimidation and killing of those who support the government or breach their extreme interpretation of the *shari'a* law', and when 'the security forces' response to their presence becomes too intense', the flying column simply moves on to another valley and engages in the same activities in a different locality.[33] Of course the Taliban retains a permanent presence in the valley and villages through the part-time local guerrillas. But they are not as important as each village's clandestine, underground cell, which sometimes in fact functions as 'a shadow government structure with the cell leader acting as the "guerrilla mayor" of the village'; in other cases the cell merely 'rallies local fighters to meet the "flying column" when it is in the valley, coordinates local support work for the full-time fighters, conducts armed propaganda, and provides leadership for local guerrillas'.[34] Through the presence of the clandestine cell, the Taliban can retain some control of the village even if it is temporarily occupied by the security forces – and can be sure of re-establishing full control when these forces leave and it is time to prepare for the next visit by a flying column of guerrillas.

The rural and urban O1 zones are also a likely location for the insurgents to find local allies and to consider investing time and effort into converting them into political supporters. Local alliances occasionally make strange bedfellows. For example, some of Mao's communist insurgents operated in the late 1930s in regions behind Japanese lines where there was relatively little insurgent military capability and they were required to 'adapt more effectively to the local environment', to the extent of sometimes making alliances with powerful gentry in order to exploit anti-Japanese patriotism and the local social pressure the gentry could exert on behalf of their communist allies.[35] The communists would have been tempted to win over these gentry allies, or even the social following of the gentry, but would have had to weigh up the downside risks of attempting some demand-side or supply-side politicisation, which might very well end a valuable alliance that had been providing them with an O1 control of the region. In comparison, the Taliban's local alliances with Pashtun tribal leaders have offered more opportunity for conversion, not only because of the Taliban's military capability but also because the Pashtun tribes are not very ordered and hierarchical, compared to Arab tribes, and therefore offer a significant social-political opportunity for conversion.[36] Local alliances have often led to the tribal leaders either being co-opted as political supporters or being subordinated (to an insurgent leader or a local religious leader co-opted by the Taliban) and seeing their social followings subjected to Taliban attempts to convert them into political supporters of the insurgency.[37]

However, insurgencies find it increasingly difficult and hazardous to win over potential political support as the military-political environment becomes less favourable. For example, a feature of areas in which the Peruvian SL insurgents of the 1980s–90s had 'only limited influence' was the absence of the continual political-education meetings that characterised life in the 'guerrilla zones'.[38] From the local population's perspective, too, it becomes increasingly difficult and hazardous to supply the insurgents with active political support as the military-political environment becomes less favourable.

These changes are very evident even in the shift from O1 to O2 levels of insurgent control. The O2 zone of shared (day/night) or alternating control is the most complex in terms of support as well as military-political control. Although this zone seems to be comparatively rare, Wood's analysis of the FMLN insurgency in 1980s El Salvador included a case study of what appears to be an O2 zone in a rural area:

> both the FMLN and government troops maintained a presence in the region, the FMLN in small encampments in the rough terrain both north and south of Tierra Blanca and the government in bases in Tierra Blanca and the nearby towns ... Occasionally, one side or the other would mount an offensive beyond their bases.[39]

When either side's forces took temporary control of an area beyond their bases, most of the inhabitants of that area behaved as fearful subjects of these rulers, whether FMLN or government forces, for everyone 'felt they had to supply food and water to combatants of either side when asked to do so'.[40] However, some of the population would behave as political supporters of the insurgents when the FMLN were temporarily in control of the area. This third of the population were categorised by Wood as supplying support (in other words, political rather than fearful-subject support) to the insurgency because this third provided 'information and supplies beyond the contribution' required of fearful subjects when the insurgents were in control but would *not* give 'information and supplies to government troops beyond the necessary contribution' required of fearful subjects when the *troops* were in control.[41] Such political support was motivated, as described in Chapter 6, not only by the insurgents' political goals but also by the moral selective incentive of pride in participating in the insurgency. Nonetheless, these political supporters were not 'suicidal' and 'did not carry out their activities without some regard for their safety', that is, some regard for their 'proximity to insurgent forces'.[42]

In the more unfavourable, state-dominated military environment of an O3 zone the insurgents will have difficulty carrying out even clandestine attempts at converting potential into actual political support. Indeed it is difficult for the insurgents even to elicit and receive support from their existing political supporters. Kalyvas argues that the insurgents may still be successful in using coercion in an O3 zone so long as they 'cultivate a perception of credible selection' in their use of violence, that is, by having local agents (clandestine cells or informers) who provide the insurgents' hit squads with accurate information about whom to punish for failing to supply at least passive support.[43] But this coercion is only preventing neutrals and perhaps hostiles from giving information to the state's forces; it does not enable political supporters to supply active support without fear of punishment from the state's forces. Some political supporters may well be willing to risk this punishment because they are influenced by moral or social selective incentives. Yet the insurgents themselves are taking risks in even eliciting such active support, such as knocking on doors asking for logistical contributions or assistance.

In the O4 zones the insurgents will face a yet more daunting prospect. They will lack any coercive capability and will be unable to rely on passive support from anybody except political supporters, who will also face even greater risks than in an O3 zone in supplying active support. The difficulties of operating in an O4 zone are illustrated by the QAP (Al Qaeda on the Arabian Peninsula) campaign against the Saudi government that was described in Chapter 4. In the initial stages of this 2003–4 campaign the QAP's active political supporters 'organised safe houses, rented cars, provided other logistics such as opening bank accounts, and formed a potential reservoir of fighters', but this support base was inevitably whittled away by the security forces' counter-offensive and arrest of many active supporters of QAP.[44] More importantly, the insurgents were unable to restore their support base by converting potential into actual political support, not even by converting passive into active supporters. The QAP had plenty of sympathisers who 'followed closely the websites of the QAP and spread its ideas' as well as supplying the passive support of denying information to the state's forces.[45] But the sympathisers were 'intimidated enough to prevent them actively supporting the QAP', for when 'state repression swung into gear the risks became too high'.[46] Obviously, the risks were also too high for the insurgents to have any prospect of restoring their support base by attracting new political supporters.

However, whatever the operational zone, from O1 to O4, the development of political support is a crucial aspect of the insurgents' overall opportunity for success. Their reliance on political supporters, not fearful subjects or local allies, in the operational zones is confirmed by Crenshaw's classic study of anti-colonial terrorism in 1950s Algeria. What she termed 'compliance terrorism' was used by the expeller FLN to ensure that the indigenous population supplied it with support. From the beginning the insurgents realised that the 'only means of preventing the population from betraying them was clandestine coercion', which was indeed successful in 'intimidating the Algerians who would have provided essential intelligence information about the FLN to the French'.[47] Then the insurgents went on to use widespread intimidation when eliciting *active* support. There was such a 'pervasive and continuous' threat of mutilation or death that ordinary Algerians 'were afraid of FLN violence' and in many cases provided the active support demanded by the insurgents, which included joining in boycotts or other political activities and assisting with sabotage or other military activities.[48] But the coercive pressure of compliance terrorism seems to have been only part of the explanation for the FLN's ability to elicit support in operational zones:

> Only the *actual* and *potential constituency* of the nationalists could be coerced into limited support. The heightened susceptibility of the Algerian audience resulted from its acute vulnerability to terrorism and from extrinsic factors [ethnic affinity and ideological sympathy] that encouraged obedience to the FLN.[49]

In other words, the active support coercively elicited in operational zones was not coming from 'fearful subjects' hostile or neutral towards the FLN but from actual or potential political supporters who were being coerced in order to overcome collective-action problems.

In fact insurgents can actually alienate potential supporters by the overuse of coercion in operational zones. The communist insurgents in Malaya began in 1948 by relying on compliance terrorism to deter the immigrant Chinese population not just from betraying the insurgents but from cooperating with British colonial rule. Within a few years the insurgency's leaders realised that its use of 'calculatedly brutal' intimidation in this unrealistic manner 'had damaged its image with the rural Chinese population'.[50] But although its leaders therefore adopted a more realistic approach, its alienation of potential support within this crucial sector of society could not be rectified before the British resettlement and 'hearts and minds' counterinsurgency strategies began to take effect.

By comparison, the FLN in Algeria adopted from the outset a more measured approach that did not expect too much of coercion and combined it with attempts to develop political support and with the use of moral and social selective incentives. Thus the potential victims of compliance terrorism apparently felt dishonour and shame as well as fear while, on the other hand, 'pride and status' could be gained by abandoning outward neutrality and becoming an avowed political supporter of the FLN.[51] The insurgents had clearly been successful in converting many potential supporters into 'rebels with a cause' – that of supporting the FLN's anti-colonial insurgency.

Ethnic/sectarian support

A special case of social opportunity and social-political environment arises when insurgents are attempting to win political support from an ethnic or sectarian group. However, this is so common that it seems a misnomer to describe it as 'special' in the sense of unusual. For in addition to being characteristic of history's many separatist insurgencies, an attempt to win ethnic/sectarian political support has sometimes occurred in expeller insurgencies and even in revolutionary insurgencies. In fact revolutionary insurgents might be the first to agree that modern insurgencies have been given greater social opportunities by ethnic/sectarian divisions than by class divisions.

Of course ethnic/sectarian social divisions have sometimes produced more costs than benefits for a revolutionary or expeller insurgency. The classic case occurred in the 1940s–50s in the British colonial territory of Malaya, where communist revolutionary insurgents were faced with an alliance between Britain and the territory's Malay ethnic (and Islamic sectarian) majority: 'the communist movement was identified with the Chinese and was therefore regarded as alien by the Malay population'.[52] More recently, after the 2003 invasion of Iraq the many bands of Iraqi expeller insurgents were faced with an alliance between the US and two of Iraq's three main ethnic/sectarian groups – the Shiite sectarian majority and the Kurdish ethnic minority.

In other cases, though, ethnic/sectarian divisions have provided greater opportunities for the insurgents than for their foreign foe. The most paradoxical of these insurgent opportunities is the 'combative culture' that is often found in countries with a history of civil conflict or hostility. As Liberman has argued, long-standing ethnic or ideological conflict hones 'organizational and fighting skills' that can be turned against foreign invaders.[53] And O'Neill pointed out that the tenacity of Afghan resistance to the Soviet foreign invaders of the 1980s can be partly attributed to the combative nature of the country's 'various ethno-religious groups', who had a long history of defending tribal autonomy and harbouring animosity toward other groups.[54] Although this history had in turn encouraged disunity between anti-Soviet expeller bands based on different social groups, it had also offered each band a social-political opportunity to establish a broad, deep and strong support base within its own social group.

The same sort of argument applies to separatist insurgencies that can win the political support of such an ethnic/sectarian minority – by supply-side responding to an existing 'cause' and/or by the demand-side creating of a 'cause'. The most famous separatist insurgent bands of recent times have been responding to an existing 'cause', as in the case of the Provisional IRA, Basque ETA, LTTE Tamil Tigers, Kurdish PKK and Eritrean EPLF, even if they have therefore had to compete with other bands or political organisations to win the political support of their aggrieved ethnic/sectarian group. However, a separatist insurgent band may well be facing the demand-side problem of a lack of discontent and grievance, for their ethnic/sectarian group may not be suffering from any relative deprivation, such as loss of privileged status, or from any kind of cultural, political or economic discrimination.[55]

A yet more fundamental demand-side problem is that the ethnic/sectarian group may even lack sufficient collective identity to be susceptible to demand-side manipulation. The group's sense of ethnic/sectarian identity may be only weakly developed or may be obscured by rival and cross-cutting sources of identity, such as regional, tribal, religious or even national identities.[56] In these cases the separatist insurgents will have to strengthen a weak ethnic/sectarian identity or increase its saliency in relation to other, rival sources of identity so that it clearly takes priority over the individual's various *other* identities.[57]

As was argued in Chapter 4, the provocation strategy seems the best method of exploiting this minimal social opportunity. If successful, the separatist insurgents will have accomplished the first step in what Byman's 'logic of ethnic terrorism' considers to be their first task and mission – to make ethnicity a salient political issue for their ethnic group.[58] But even if the insurgents have created an ethnic/sectarian identity and political grievance, they will still have to overcome the supply-side problem of selling their political goals as a collective good that will satisfy the grievance. Their success will depend largely on whether or not the social-political environment contains a particular geographical factor and opportunity, namely whether or not their ethnic/sectarian group is concentrated in a particular territorial region of the country:

> Ethnic groups will seek to rule territory if they are geographically concentrated in a particular region of a country, especially if that region is a historic homeland. They will show little interest in controlling territory when they are either widely dispersed across the state or concentrated only in cities.[59]

Furthermore, Toft argued that in addition to the geographical factor, there were two preconditions for an ethnic group to demand sovereignty over territory. First, the group 'must believe that their cause is legitimate' and a 'just cause', which is the sort of belief that any insurgency would hope to instil during a demand-side politicisation.[60] Second, the group must believe their capabilities 'give them a reasonable chance of gaining control of the territory they desire' – that they have the 'capacity to wage a successful fight for independence'.[61] This is the sort of belief that an insurgency is *not* likely to have instilled in the demand-side politicisation, and any insurgency that has been given the geographical-factor opportunity will still have to surmount this capability hurdle at the supply-side step of converting potential support into actual political support. If the ethnic/sectarian group will not buy the idea of seeking the full sovereignty and independence of secession, the insurgents may have to adjust their political goals to merely seeking a federal or other form of autonomous control over territory.

Another special feature of seeking ethnic/sectarian political support is that it may also involve using a special sort of social selective incentive. In Putnam's famous theory of social capital he pointed to how ethnic groups can be socially isolated or 'segregated' within a much larger population:

> Kinship ties have a special role in the resolution of dilemmas of collective action. In some respects bonds of blood are comparable to horizontal ties of civic engagement, but family is more nearly universal. It is no accident that family firms *and close-knit ethnic minorities* (Jews in Europe, overseas Chinese in Asia, and so on) have been important in the early stages of the commercial revolution. ... Dense but *segregated* horizontal networks sustain cooperation within each group.[62]

In addition to these cooperative relationships, there is a normative form of social capital that dramatically extends in-group cooperation to include people who are not an individual's family, friends or familiars. What Putnam terms the 'norms of generalized reciprocity' enable 'bonds of blood' to operate in *extended* kinship groups, in which there may be little face-to-face familiarity with distant relations.[63] An obvious example is the way that vendetta norms of reciprocity may overcome the collective-action problems of combining to avenge the death of a distant and unfamiliar kinsman. Similarly, an individual may apply norms of reciprocity to any and all members of his large and largely unknown ethnic/sectarian group, as in Putnam's examples of European Jews or overseas Chinese. And these norms may be enforced not only by moral selective incentives but also by the social selective incentives deployed by the individual's local kinship and peer group.

These sorts of social selective incentive may well have been operating in the case, mentioned in Chapter 8, of financial support being delivered to the LTTE Tamil Tigers by ethnic Tamils living overseas and even in far-off Western countries. They may well be operating, too, in some cases of sectarian political support being given to not separatists but jihadists or Islamic revolutionaries by members of Islamic immigrant communities in Western countries.[64] However, in both these immigrant situations the personal-identity form of moral selective incentives is likely to be operating. When described in Chapter 6, this form of moral selective incentive was depicted as occurring in ethnic/sectarian groups, and it is just as likely to be operating when an individual is merely supplying passive or active support to a band rather than actually joining it. Similarly, the individual may be motivated to supply support by identity-based emotional loyalty, which is an expanded version of the affective form of selective incentive that was also described in Chapter 6. Social psychologists have pointed to emotional identification with ethnic and national groups, and presumably could also have pointed to identification with sectarian groups, as examples of how group identification 'means expanding the boundaries of what we care about' and may include 'groups that are so large that the great majority of members are personally unknown to us'.[65]

However, insurgents will usually derive most of their social support from within their ethnic/sectarian group's home territory rather than from among members of the group who are living in other countries. In some insurgencies the band may have established semistate, O1 or O2 levels of control over parts of that territory, as the Tamil Tigers did in Sri Lanka from the 1980s to 2009. And in those base areas or operational zones they will be able to use similar selective incentives to those used by revolutionary or expeller insurgents, such as the social selective incentives available in many rural communities.

Even in urban areas they may be able to use an equivalent of these village communities. Calhoun pointed out that 'urban ethnic groups should be at least as important as rural villages in our images of community' even though urban ethnic communities tend not to be as small, densely knit and autonomous as traditional peasant or artisan communities.[66] Nor will they be socially segregated urban communities if they belong to a geographically concentrated ethnic/sectarian group that constitutes a regional majority. And certainly the urban areas of separatists' semi-states or O1–O2 zones will likely be predominantly inhabited by their ethnic/sectarian group. However, some such groups have intra-ethnic or intra-sect *sub*groups, such as tribes or clans, that are found in urban as well as rural areas and may provide a relatively small and densely knit urban community. The best-known examples are the clans that have dominated politics in urban areas of Somalia and the tribal or clan groupings that still influence many Sunnis and Shiites in urban Iraq.[67] But in the urban areas of most modern societies it has more often been revolutionary than separatist insurgencies that have been seeking to convert potential into actual political support and perhaps even insurrectionary political support.

Urban and insurrectionary support

Revolutionary insurgents' social opportunity to win political support in urban areas has been the subject of much debate. So it is safer to focus instead on their social opportunity to use social selective incentives to overcome collective-action problems. Calhoun has argued that when industrialisation replaced artisan workers and communities with a modern working class and formal labour organisations, it also replaced artisans' form of rebellious collective action with a weaker form. Protests became 'shorter, less sustained and concerted efforts' without the 'unifying social basis for collective action which community structure provided'.[68] Waterman, however, argued that the strong social ties required to mobilise strong collective action can be forged in industrial workers' subcultures – by labour organisations which seek to tie workers to each other as well as to a political movement.[69] But if an insurgency seeks to create such worker-subcultures, it will usually discover that this is a much more risky investment of time and effort than seeking to organise the peasantry for collective action. The state will probably have established an O3 or O4 zone in its major cities, and even in some O2 zones it would be very difficult as well as dangerous to create labour organisations that will tie workers to each other and to the insurgency as a political movement.

Of course it would be less difficult and dangerous to take over an *existing* labour organisation that has already done the hard work of creating a worker subculture. But it is unlikely that the social-political environment will offer any opportunity for taking over such organisations. They will already be tied to a political movement and so either will have to be persuaded – through some brilliant supply-side marketing – to switch their political support to the insurgents or will have to be presented with an alliance that offers them the best prospect of achieving their organisations' goals. An alliance between insurgents and Peronist labour unions seems to have occurred in Argentina in the early 1970s but its eventual dissolution (and the Tupamaros' inability to establish an alliance with the Uruguayan labour unions) indicates how difficult it is for even relatively successful and militarily capable insurgents to exploit this social opportunity.

Furthermore, the social-political environment will probably not be offering the insurgents any opportunity to use the typically urban manner of overthrowing the state – namely, mass insurrection. Such insurrections require a strong rebel cause as well as strong social or moral pressure to produce a large number of very active and dedicated supporters who are willing to risk being arrested, assaulted or even gunned down by the state's forces. Indeed it is such an active form of support that the line between support base and insurgent band will be momentarily blurred or even transcended, as the supporters might well have to engage in the insurrectionary equivalent of combat operations.

So it is not surprising that historically the working class has proved reluctant to risk an insurrection unless and until there is a change in the military-political environment that reduces the danger of state repression. As Waterman pointed out, even the most rebellious and spontaneous of uprisings, the revolutionary 'crowd' or 'mob', seems to have arisen only when 'the risks appeared to be relatively low, or the prospects were tested in a gingerly fashion before large numbers of people took the plunge'.[70] This is also true of the spontaneous street demonstrations by peacefully protesting crowds that have sought to topple unpopular non-democratic regimes and replace them with democracies.

But while these demonstrators have had political grievances and goals, it is hard to see how they overcame their collective-action problems – especially as they were not engaged in emotional mob-like violence but in peaceful street demonstrations. One likely explanation emerged after ten years of successful mass protests in a number of countries, ranging from the Iranian revolution in 1978–9 to the overthrow of communism in East Germany and Czechoslovakia in 1989. Kuran's explanation was that a chain reaction or bandwagon effect had occurred once some morally pressured individuals began to ignore the risk of repression and to express publicly their previously concealed preferences for a change in regime.[71] When their public protests were not repressed by punitive countermeasures, other discontented and politically aggrieved but more cautious people began joining the protests in ever-increasing numbers (and were probably responding to social as well as moral incentives). Consequently, the presence of a *latent bandwagon* had meant that some event which otherwise would have had little effect could instead trigger a revolution – in fact a single protesting individual can 'set off a chain reaction that generates the necessary mobilization'.[72]

However, the process seems to have operated differently in Czechoslovakia than in East Germany. In the case of what became known as the 'velvet revolution' in Czechoslovakia, it appears to have been particular segments of society who got the bandwagon moving, with a series of almost exponentially increasing demonstrations in November 1989 that culminated in a massive general strike. Petersen's analysis of the Czechoslovakian case acknowledges that the East German case differed in spreading so rapidly across different social groups, 'as a flood, rather than a cascade', as well as not including an organised strike and involving only street demonstrations.[73] In comparison, his model of the Czechoslovakian case is of a three-stage process: 1) dissidents begin the process because they have already risked and suffered regime sanctions; 2) if dissidents' protests are not brutally repressed, this sends a signal about demonstrating to students, who have more to lose than dissidents because they have their future careers to consider; and 3) if students are not deterred or stopped by the regime, this in turn sends a signal about strike action to industrial workers, who have more to lose than students because they can lose their jobs and often have families to worry about.[74]

In both these cases an insurgent band might have provided the trigger that transforms a latent bandwagon into a mass insurrection by protesting crowds. Kuran recognised the role that might be played by revolutionary leaders and he acknowledged that an organised opposition group 'may have a tremendous impact' by pushing 'a bandwagon into motion'.[75] In other words, an insurgent band may be able to defeat the state by eliciting support not for their insurgency but for a peaceful street demonstration against an unpopular regime, which gets the bandwagon moving and diverts the regime's energies into coping with a democratising insurrection. In fact the insurgency's 'supporters' may not even realise that they are being used by the insurgent band for its own, revolutionary ends rather than for meeting the popular demand for democratisation. Yet although the presence of a latent bandwagon is the most dramatic opportunity offered by any social-political environment, the insurgents may not be well placed to take the opportunity or may not even realise that it has appeared – as Kuran emphasised, the bandwagon insurrections have often been 'unanticipated' by friend and foe.

A similarly unanticipated opportunity can occur in more orthodox cases of urban support and insurrection. A well-known example is the huge expansion of the Sandinistas' urban support base in the year preceding their final offensive of May 1979 against Somoza's regime, thanks to an inadvertent provocation strategy and the nature of the regime (see Chapters 4 and 8). This seems to have been an unexpected opportunity for the Sandinistas, which they exploited by rapidly increasing their political support and local alliances in the urban areas. They 'worked assiduously to channel the manifest hostility to Somoza', not only organising support groups amongst the urban poor but also encouraging the formation of Sandinista-allied umbrella coalitions of moderate, including middle-class, organisations opposed to Somoza's dictatorship.[76] By skilfully exploiting the ready 'cause' of overthrowing the dictatorship, the insurgents were able to transform the broad and strong hostility to Somoza into a broad, deep and strong urban support base that would become insurrectionary when a military opportunity was created by the Sandinista rural military offensives. The character of the Nicaraguan revolution in this final phase 'is reflected in the social background of the rebel combatants killed during the insurrection, most of whom were students and tradespeople (artisans, food vendors, and the like)'.[77] These casualty lists also reflect the fact that rising up against a state has always been dangerous, whether that state is the Roman Empire or a neopatrimonial dictatorship in Central America.

Conclusion

Summing up

In the 21st century the world is witnessing the dangerous anomaly of anachronistic forms of rule and warfare, namely non-democratic rule and stateless warfare, continuing to survive and even flourish thanks to their ability to modernise themselves. In Part I it was shown how stateless warfare, like non-democratic rule, had undergone three phases of modernisation, which had enabled the insurgent variant of stateless warfare to continue playing an important role in modern history. These three historical phases of modernisation had begun some two centuries ago with a new emphasis on nationalism that has superseded the localist and religious orientation of premodern insurgencies. The second phase, introduced by Mao Zedong, provided revolutionary insurgent bands with a new approach, namely establishing semistate base areas in which the insurgents could create regular armies capable of destroying the local state in a revolutionary civil war. But the third phase of modernisation has produced insurgencies that prefer to rely on new techniques and technology rather than on the maoist model's approach of shifting to semistate warfare.

The chapters of Part I have also examined the political goals of insurgent bands and how they use organised violence in their attempt to achieve these goals. In addition to having experienced three phases of modernisation, stateless warfare has developed three main types of insurgency – the expeller, the separatist and the revolutionary – that are related to the different types of political goals that insurgents pursue: expelling a foreign state; separating from the local state and establishing a new, independent state; or destroying the local state and establishing a revolutionary new form of state and society. The organised violence used by the insurgents in their attempt to achieve these goals has in turn been analysed under the three headings of mode, format and strategy. More specifically, these three aspects of how they use violence are 1) the guerrilla or terrorist mode of operation, 2) the bandit-like or gangster-like operational format and 3) the particular strategy used, whether a) the maoist-model preparation for semistate warfare or b) the pressure strategy of indirectly exerting political pressure on the enemy state or c) one or both of the typically terrorist strategies of propaganda and provocation.

In Part II the focus has been on providing an analytical framework for explaining and predicting the existence of insurgencies and particularly their degree of success. As was explained at the end of Chapter 5, this 'calculus of insurgency' is based on the forensic calculus of intervention/usurpation that has long been used to explain and predict the occurrence of military or one-party dictatorships. The chapters in Part II explored the three variables in this forensic calculus of motive, means and opportunity. However, the methods aspect of the means variable had already been covered in Part I's analysis of strategy, format and mode. So the analysis of the means of insurgency has been focused on the capability aspect, not only the Personnel, Information and Equipment factors but also the quality of the band's leadership – the situational awareness of its leader and his ability to adapt the band's methods to its local environment and to changes in that environment.

The way that the local environment changes from place to place and time to time is also a crucial factor in assessing the opportunity variable. Chapter 8 explored variations in the local environment's physical, political and military dimensions as part of the chapter's analysis of the insurgents' physical, political and military opportunity. The variations in physical opportunity and environment included the difference between rural and urban environments as well as differences within them. The political opportunity was seen to differ according to the type of insurgency, with democracy removing the opportunity for revolutionary but not separatist insurgents, and to extend into the political environment of a foreign state if an expeller insurgency is using a pressure strategy against it.

However, the military opportunity was the most wide-ranging and complex. For example, it included not only the counterinsurgency method and capability of the insurgents' enemy state but also the insurgents' very-local military opportunity within the various types of operational zones. These very local military-political environments were categorised into four types of operational zones, namely the O1, O2, O3 and O4 zones, on the basis of how much control is exercised within the zone by the insurgent band or the enemy state. And the operational zones involved another facet of insurgents' military opportunity, in this case exploiting their degree of military-political control over the local population to extract support from 'fearful subjects' or elicit support from 'local allies'. This informational and logistical support in turn represents another form of military opportunity (and capability) for insurgents and is a reason why they will invest time, effort and risk in boosting their social support base, most notably in converting potential political support into actual support. It is more resilient than the support supplied by fearful subjects or local allies and will be supplied even in operational zones where the state exercises more control than the insurgents.

In fact the final chapter focused on how the social-political environment offers insurgents a varying amount of social opportunity to convert potential into actual political support. It has examined insurgents' ability to exploit and even create political grievances in favourable operational zones and in the special cases of ethnic/sectarian and urban support. And, like the analysis of motive in Chapter 6, it has shown how grievance-based support for the insurgency's political goals must be combined with selective incentives – coercive, social, moral or some other form – in order to motivate an individual to contribute to the collective action of supplying support to an insurgency or actually joining an insurgent band.

In a sense the analysis of insurgents' motive for joining an insurgency is the place where the final calculus of insurgency is indeed 'calculated'. This is because the motives for joining an insurgent band are counterbalanced by a) inhibiting motives that include b) a fear of failure that in turn includes c) a subjective assessment by the potential participant of the prospects for tactical or strategic success. The question of tactical failure or success is assessed largely in terms of the operational zone in which the recruit is likely to be fighting, especially the military opportunity and capability available to the insurgents. But the question of the degree of strategic success that might be achieved by the insurgency is a much wider-ranging and complex question and involves applying the whole insurgency calculus to the whole local environment of the insurgency.

Applying the whole insurgency calculus in a subjective assessment of the likely degree of success requires a series of strengths-versus-weaknesses net assessments. First, there is the net assessment within the means variable, including both a) method and b) capability's PIE factors and quality of leadership. Second, there is the net assessment within the opportunity variable, including the physical, political, military and social opportunities. Third, there is the key net assessment of means versus opportunity, which assesses whether the strength of the insurgents' means is counterbalanced and outweighed by the weakness of their opportunity and vice versa whether the weakness of their means of insurgency is counterbalanced and outweighed by the strength of their opportunity. If the two variables are calculated to be present in sufficient net quantity to give some hope of success, then the insurgent's fear of failure will have been assuaged (depending upon the individual's aversion to risk) and it will not be an inhibiting motive.

Much the same calculation would be made in any academic objective assessment of the insurgency's prospects. But this assessment would pay more attention to the motives of the insurgents than merely viewing them as the motivation element in the quality aspect of the Personnel factor of the means variable. Furthermore, academic assessments take a historically and geographically broader view, seeking to predict the fate of other insurgencies or potential insurgencies around the globe and seeking to explain the fate of past insurgencies throughout modern history and even further back in time. In turn this broader historical perspective begs the question of what will be the future history of insurgent stateless warfare – of what lies ahead for the next generation of insurgents.

The shape of things to come

The title of this section is borrowed from the Postscript of the prophetical analysis of future war that van Creveld put forward at the end of the Cold War. His argument that conventional, inter-state war would be replaced by something different in its form and in its participants was obviously correct but used a very broad definition of war as 'purposeful, organized large-scale violence' and used a broad definition of the non-state groups engaging in this violence:

> In the future, war will not be waged by armies but by groups whom we today call terrorists, guerrillas, bandits and robbers, but who will undoubtedly hit on more formal titles to describe themselves. ... While rooted in a 'population base' of some sort, that population probably will not be clearly separable either from its immediate neighbors or from those, always the minority, by whom most of the active fighting is done. A warmaking entity of any size will have to be 'in control' of a territorial base of some sort.[1]

His concluding remark about controlling a territorial base confirms that he was referring to semistate as well as stateless warfare but he pointed out nearly a decade later that the lack of a territorial base gives stateless bands a major advantage: 'since they do not own sovereign territory and consequently cannot be threatened with nuclear annihilation, they will able to fight each other, and the state, to their heart's content'.[2]

Deterrence is so problematic in stateless warfare not only because the foe lacks a territory and population to lay waste with nuclear or conventional weapons but also because the foe may have no concern for his people's or his own survival. After all, even warfare between states has sometimes been bedevilled by political leaders who believe they have a personal mission that transcends such conventional concerns. The obvious example is Hitler's belief that only his supposed political genius could turn history 'onto new paths' and that his mission involved foreign policy goals that 'made Nazi Germany a rogue state. Hitler recognized the brutal alternatives on which his foreign policy rested – *Weltmacht oder Niedergang* (world power or defeat)'.[3]

In fact the prospective insurgent forces of the 21st century have been described as the equivalents of Hitler's Germany and other 'have not' powers of the 1930s. Hirst pointed to their similar 'determined will-to-power based on violent action' and also warned that they will be more numerous and diverse than any 'have not' or 'rogue' states:

> Many non-state forces will engage in armed struggle both against the elites in the Third World countries and against the advanced countries. Their causes will vary from deep-green environmentalism, to religious fundamentalism, to nationalist secessionism, and there will even be some socialist revolutionaries. Such groups are likely to proliferate, acting across states.[4]

But it is always easier to predict the future path of technological developments than the future path of such political developments as the nature of insurgents' political grievances or goals. For example, in 1909 it would have been very difficult to predict that in 1939 the world would be entering a global war that included fascist and communist ideological one-party states, but it would have been much less difficult to predict that the new technology of submarines and flying machines would play a key role in such a world war. Similarly, it is relatively easy to predict that in the decades ahead the third-phase insurgencies' distinctive emphasis on the use of new techniques and technology will be taken a step further – into the area of apocalyptic technology and techniques – and thereby usher in the fourth phase in the modernisation of stateless warfare.

For within a generation there will likely be a 'revolution in military affairs' (RMA) that will give insurgent bands in the 2030s a greater military power to hurt than was available to 'have not' or 'rogue' states in the 1930s. An assessment of the frequency of RMAs in Western history identified at least five previous instances of such military revolutions, with the latest being the advent of nuclear weapons and ballistic missiles and then the application of computer and other technology to command, reconnaissance and the use of conventional weapons.[5] The sixth revolution in military affairs will take some of these developments another step further and vastly increase the potential military capability of insurgent bands. It is the 'miniaturisation' revolution in the sense of being based on a technological revolution in computers, biology and nanotechnology that is characterised by miniaturisation and in the sense of providing small groups – and even individuals – with a military power to hurt that was once the prerogative of only major states or even superpowers. Such non-state groups as insurgent bands will be capable of inflicting the same apocalyptic degree of hurt upon their foe's society as can now be inflicted by a nuclear-armed submarine.

The prospect of an insurgent-favouring military revolution can be inferred from the existing and future developments in three types of weapon technology – computers and cyber warfare, genetic engineering and biological warfare, nanotechnology and the new form of warfare that it is likely to produce. What has been described as 'knowledge-enabled mass destruction' (KMD), based upon the 21st century technologies of genetics and nanotechnology, was also described as being within the reach of small groups and even of individuals.[6] So it is within the reach of microbands and Unabomber-type individual insurgents within Western societies or the Third World. And the same could be said of a 20th century technology that has already produced the potential for KMD insurgency. Several years ago van Creveld described information warfare in cyberspace as a potential 'revolution in military affairs' that would be most advantageously used by non-state organisations because they were 'more mobile and less strongly bound to one place':

> In informational war, both geographical space and time are irrelevant. Attacks hardly require a base. ... [The prospects of] using a laptop to single-handedly bring down the strategic information structure of a big power are widely exaggerated. Still, as experience has shown, in the absence of adequate defenses even the damage

> that a lonely hacker working his laptop through a telephone link located at the North Pole can do may be enormous. Provided it is aimed at the right target and in the right manner, it could even be catastrophic.[7]

This may be the most important way in which the Internet has created new opportunities for insurgents, even if there are emerging developments that may prove almost as significant in the long run and in a less directly military fashion.[8] But it is by no means the only way in which new weapons technology is likely to produce an insurgent-friendly military revolution.

Biological warfare, too, will contribute to this revolution, largely through the development of technology for genetically altering viruses and bacteria. A few years ago it was noted that a group like Al Qaeda would have more difficulty acquiring a nuclear weapon than 'genetically altered biological weapons. The price tag is very high, but the black market for such products exists', and any 'terrorist willing to die of genetically altered and vaccine-resistant smallpox can kill literally hundreds of thousands of people'.[9] Looking further ahead, Kellman's recent account of 'bioviolence' noted that the emerging technology of synthetic genomics offers the prospect of not only 'programming' viruses to do particular things but also 'creating altogether new pathogens'.[10] The rapid spread of the naturally created H1N1 virus provided a recent illustration of the power to hurt that would be available to any insurgents who were capable of letting loose a cocktail of designer-viruses.

Nanotechnology's ability to create molecular-scale machinery – and weaponry – is the third technology that will contribute to the new military revolution. At the beginning of the century Hirst estimated that it would take at least 30 years for nanotechnology to reach its full potential in terms of the development of 'microweapons and nanoweapons', but he also noted that 'microweapons may not be that difficult for other powers to develop'.[11] So even if this weapons technology is still quite new in the 2030s, it is likely to be readily available to insurgent bands. And the three prongs of the military revolution are then likely to produce technological or operational synergies as well as giving insurgent bands the capability of attacking in three different ways against foes who must spread their defensive efforts across three different technological and operational 'fronts'.

The new capability will also affect the methods that insurgents use, with a tendency to pursue the pressure strategy with a gangster-like format and a terrorist mode of operations. The threat of using apocalyptic weapons enables the insurgents to exert huge political pressure through their potential power to hurt and indeed the fourth-phase insurgencies fighting to expel a foreign state have a much more favourable political opportunity – and much better chance of achieving a degree of success – than have their third-phase counterparts. What is more, these insurgencies may not need much of a support base, the typical O4-zone base may be sufficient, because they are deploying weaponry that may be available to even a microband if it has access to the necessary personnel and equipment.

However, unlike the expeller type, the separatist type of insurgency will need a deal of political support if it is to achieve its political goals. Obviously it needs sufficient political support from its ethnic/sectarian group for some form of separatism to be feasible in practice. For if most of the group are not interested in secession or autonomy, it is hard to see how sovereign or devolved political institutions could actually operate after they were conceded to the insurgents by the local state.

The revolutionary type is in an even more paradoxical position. On the one hand, it has the advantage of no longer being constrained by the lack of political opportunity in democracies or semidemocracies that was described in Chapter 8, for it no longer has to develop much of a revolutionary movement in order to have an impact. On the other hand, it has little more political opportunity than its separatist equivalent of achieving its political goals if it lacks sufficient political support. Although it may be able to pressure the local state into radically changing the society and state, it is unlikely that these state-implemented changes will accomplish the revolutionaries' goals unless they are politically supported by a substantial amount of the population.

In fact it is likely that some revolutionaries will opt for a different and dramatically new insurgent strategy, which will seek to destroy the state and much of the fabric of society by actually *using* apocalyptic weapons. For example, new versions of the anti-modern microband insurgency pioneered by the Unabomber may employ this strategy. Instead of retiring to a cabin in the woods, the new opponents of technological society may seek to use its tools against it – working away in laboratories to transform these technological tools into new weapons that are far more deadly than the Unabomber's parcel bombs. The same sort of truly apocalyptic strategy may well be used by religious extremists who believe they are the instrument of the 'wrath of God' against a materialist, godless society. As this would be a religious insurgency, it might be categorised as premodern but the key point is that its fourth-phase weaponry gives it the military capability to reduce its whole society to a premodern status.

What is likely to be done?

Although states may well want to take some drastic countermeasures against the threat of fourth-phase, apocalyptic insurgency, there are good reasons to expect that they will be unable to implement such measures. For example, by extrapolating present tendencies into the future it can be predicted that Western states will respond with an enlarged version of the 'war on terror' against anti-Western expeller insurgencies operating from locations in the Third World, especially if the locations are ineffectively policed by weak or collapsed states.[12] But in carrying out this preemptive counter-offensive the West will face a time problem, a numbers problem, and a power problem. The time problem is simply that each band of insurgents will have to be dealt with quickly, before they have the time to acquire and deliver fourth-phase weaponry. The numbers problem arises from the sheer number of bands or potential bands and was identified by Simmel over a hundred years ago: when dealing with a diffused multitude of weak enemies rather than a single strong foe, 'one may oftener gain isolated victories, but it is very hard to arrive at decisive results which definitely fix the relationships of the contestants'.[13] The power problem is that Western societies lack the military power to exercise control over the Third World territories that are not under the control of their own governments and therefore are safe havens for anti-Western expeller insurgent bands. As Kalyvas has pointed out, a 'staggering' number of troops are required to establish and preserve full control 'over the entire territory of a country'.[14] And the recent conflicts in Iraq and Afghanistan have shown that the US and its allies simply do not have sufficient troops available to establish and preserve control over more than a small proportion of the Third World.

Some 30 years ago Waltz's neo-realist theory of international politics pointed out the limits of military power in dealing with problems of control within a country. He noted that domestic or foreign military forces 'are insufficient for the task of pacification, the more so if the country is rent by faction and if its people are politically engaged and active'.[15] In fact he drew the distinction between 'establishing control over a territory' and 'exercising control within it', and described military force as being preeminently a means of affecting another state's external behaviour – not of affecting behaviour *within* that state's territory.[16]

Of course it be will argued that military forces *can* exercise that degree of control within states' territory if they use advanced surveillance technology to compensate for their insufficient manpower. But even Orwell's vision in his novel *1984* of the ultimate surveillance and indoctrination system did not envisage the system being applied to more than a minority of the population, notably the intellectuals, administrators and security forces.[17] After all, the two-way 'telescreens' that maintained constant surveillance and propaganda, even in the privacy of people's own homes, had to be monitored by large numbers of skilled personnel – computers could not be relied upon for this job. Consequently, the mass of the population was controlled by more old-fashioned methods which would have neutralised the masses as a political threat, as the source of an insurrection or large-scale insurgency, but could not have prevented the creation of microbands. As some future microbands will likely be capable of developing fourth-phase weaponry, the Orwellian system would therefore not be able to provide the sort of control required to counter the threat of fourth-phase insurgency. States countering expeller insurgents in the Third World or countering domestic threats from revolutionary or separatist insurgents may want to use Orwellian measures but they will be unable to implement them to the degree that would be required to neutralise the fourth-phase threat.

However, the world's many dictatorships, disguised dictatorships and semidemocracies are likely to use the fourth-phase insurgent threat as a political opportunity to implement Orwellian measures against *political* threats to the regime. Huntington warned of the possibility of an Orwellian dictatorship in his early 1990s work on history's third wave of democratisation. Although he was writing during the apparent peak of that third wave, he foresaw a reverse wave that would shift the world back towards authoritarianism and might spawn a new, technocratic electronic form of dictatorship 'in which authoritarian rule was legitimated by and made possible by the ability to manipulate information, the media, and sophisticated means of communication'.[18] He could not foresee of course how this new form of dictatorship would come about but, somewhat paradoxically, the likely manner in which it will occur was described by Huntington in the 1950s when critiquing Lasswell's theory of the 'garrison state' or 'garrison-prison state'. Huntington had been sceptical of the notion that the demcracies would evolve into dictatorial regimes and societies focused on a perpetual preparation for war and ruled by the military and police behind merely ceremonial democratic institutions.[19] But if his discussion of this possibility is reframed in terms of fourth-phase insurgency and of non-democratic regimes, then it seems a prophetic insight into how the garrison-prison state might emerge in the 21st century as the justification for an electronic modernisation of dictatorship – the fourth phase in the long history of non-democratic rule's modernisation. The perpetual warring or preparation for war against fourth-phase insurgents will be more of an excuse than a reason for this new modernisation of dictatorship. Nonetheless, there will be a sufficient link between its emergence and the earlier emergence of fourth-phase insurgencies to argue that the two historical paths of modernisation have become intertwined – that the fourth phase in the modernisation of stateless warfare has produced a fourth phase in the modernisation of non-democratic rule.

A more important issue, however, than the shift from one type of non-democracy to another is whether there will also be a shift from democracy to the new type of non-democracy. And this is what Huntington had in mind when he mentioned the likelihood of a reverse wave that might include the emergence of this new type of dictatorship. But such a huge shift in the form of regime, from democratic to non-democratic, could be carried through only by politicians with an entrepreneurial-style, pioneering character that is now seldom found in democracies and will be even rarer in future decades.[20] What is much more likely to occur is the evolution of democracies into a paranoid form of semidemocracy, in which the right to liberty is given less priority than protecting the rights to life and the pursuit of happiness. The shift to semidemocracy would not be merely temporary, as it was during the Second World War, but instead a permanent response to the permanent state of war or threat of war that has been created by the modernisation of stateless warfare. However, democracy has survived greater dangers, especially in the 1930s–40s, and doubtless a new generation will find new solutions to the new problems of the 21st century.

Notes

Chapter 1

1 Clausewitz, 1976 [1832]: 75.
2 Ibid.: 87. My emphasis.
3 Gray, 2005: 30, 342.
4 Ibid.: 217.
5 Eldredge, 1995: 49.
6 Ferrill, 1985: 20–2.
7 Keeley, 1996.
8 Ibid.: 175, 176.
9 Asprey, 1975: 3.
10 Hengel, 1975: 39.
11 Ibid.: 56, 58. The conquest of the independent Jewish state by a Roman army in 63 BCE had been followed by continual rebellions, then the rule of Herod as a puppet 'king of the Jews' from 37 to 4 BCE, another round of rebellions after Herod's death and succession by his son Archelaus, and finally the removal of Archelaus by the Roman Emperor Augustus in the year 6 and the conversion of Judea into a Roman province under the rule of a Roman governor appointed by the Emperor.
12 Ibid.: 58–9.
13 Marten, 2007: 48 and 48–50 on the warlordism of the feudal lords of medieval Europe.
14 Tilly, 1985: 174.
15 See Weber, 1947 [1920]: 154 where he defines a compulsory political association as enforcing its order within a given territorial area and specifies that it 'will be called a "state" if and in so far as its administrative staff successfully upholds a claim to the *monopoly* of the *legitimate* use of physical force in the enforcement of its order'; and see Weber, 1970 [1919]: 78 where he notes that in modern times 'we have to say that a state is a human community that (successfully) claims the *monopoly of the legitimate use of physical force* within a given territory'.
16 Asprey, 1975: 91.
17 Wright, 1964 [1942]: 51–2.
18 Ibid.: 52.
19 Tone, 1994: 71.
20 Quoted by Chartrand, 2004: 53.
21 Ibid.: 49.
22 Tone, 1994: 71, 72; Bell, 2007: 287.
23 O'Brian, 2003: 1, 16, 17, 52.
24 Bell, 2007: 287.
25 Clausewitz, 1976 [1832]: 479.
26 Ibid.: 479, 483, 480, 482.
27 Asprey, 1975: xiii.

28 Wright, 1964: 356. My emphasis.
29 See Zamoyski, 2008: 7, 8, 13, 68, 70, 102 on the leaders' ambitions; 31, 59 on motivating the troops; and 57, 69 on seeking support from Polish working class and peasantry.
30 Porch, 2006: 140.
31 Ibid.: 147, 148, 202.
32 Ibid.: 204.
33 Mao, 1966b [1938]: 182.
34 Ibid.: 171.
35 Trotsky, 1934: 225.
36 Wright, 1964: 142.
37 Ibid.: 52.
38 Townshend, 1979: 340, 343.
39 See Brooker, 1995 and 2009.
40 Wright, 1964: 51–2.
41 Schelling, 1966: 2, 3, 12.
42 Ibid.: 33. He drew a comparison between the military power to hurt and Wright's notion of the 'nuisance value' of war or the threat of war when used by unscrupulous governments as an aid to their diplomacy. But Schelling pointed out that 'we need a stronger term' – namely, the power to hurt – and 'need to recognize that even scrupulous governments often have little else to rely on militarily' and 'that the power to hurt has been, throughout history, a fundamental character of military force and fundamental to the diplomacy based upon it' (ibid.: 33).
43 Ibid.: 7.
44 Ibid.: 24.
45 This adaptation is particularly striking when an insurgency has shifted from urban to rural environment, as with the Chechen separatists after the loss of Grozny or with Al Qaeda's contribution to the Taliban-dominated insurgent 'coalition' in Afghanistan.
46 Gray, 2005: 224.
47 White, 2003: 240; Karmon, 2001: 35.
48 Lawrence, 2005: 58–62.

Chapter 2

1 Chaliand labelled them: 1) the national liberation movements, 2) the revolutionary struggles and 3) the struggles waged by minorities with secessionist or 'lesser ambitions' (Chaliand, 1982: 11). Laqueur described them as 'the three main species of guerrilla wars' : 1) those 'directed against foreign occupants', 2) those favoured by 'separatist, minority movements' and 3) those directed 'against native incumbents' (Laqueur, 1977: 395).
2 Bell, 2007: 287. My emphasis.
3 Clapham, 1998: 7. He draws attention to the comparison between contemporary Africa and the economics of warfare and plunder in Europe from the early medieval period onwards (ibid.: 6 n. 7).
4 The feudal separatist feature of being focused on winning local rather than central state power has become even more prominent in recent years in the

African context of weak or collapsed states, and indeed taking over such states is 'no longer the primary objective of insurgency' (Boas and Dunn, 2007b: 21, 23). Englebert has argued that at local level there is much for insurgents to gain from taking over 'provincial governments, regional bureaucratic agencies, parastatals, or recognized chiefdoms' in order 'to issue (however arbitrary) regulations and make extractive demands' (Englebert, 2007: 60, 61). Together with the refusal of the international community to accept any secession from African states, this has led to 'the relative scarcity of separatist motive' among African insurgencies *if* 'separatism' is defined in terms of secession rather than local autonomy, as the author did in an earlier article on 'Africa's secessionist deficit' (ibid.: 55 and 67–8).

5 This is evident in the regional distribution of the 127 civil wars (or insurgencies) identified by Fearon and Laitin as meeting their criteria in 1945–1999 (Fearon and Laitin, 2003: 77).

6 For Kilcullen contends that the Taliban should be seen as primarily 'a loose alliance of Pashtun nationalists, dispossessed tribes, and Islamist extremists that seeks to control the Pashtun-majority parts of both Afghanistan and Pakistan' (Kilcullen, 2009: 111).

7 Laqueur, 1977: 401.

8 Lawrence, 2005: 61.

9 Schmid's often-cited study of definitions of 'terrorism' not only identified more than a hundred different definitions but also could incorporate only 13 of the 22 elements commonly found in them when he sought to construct a synthesis or conglomerated definition (Schmid, 1983: 19–58, 70–111). As he drew upon writings from the period 1936–81, it is quite likely that 'terrorism' is by now the most variously defined political term in the English language.

10 Crenshaw, 1995.

11 Kilcullen, 2009.

12 Asprey, 1975: xi.

13 O'Neill, 1990: 30 n. 17, 26.

14 Wright's account of the Cuban revolution's effects on Latin America in the 1960s–80s devotes separate chapters to urban and rural guerrilla warfare (Wright, 1991: chs 5 and 6).

15 Laqueur, 1977: 321, 403. At the end of his book he suggested that the age of the guerrilla was drawing to a close with the 'retreat into urban terror, noisy but politically ineffective' (ibid.: 409).

16 Laqueur, 1987 [1977]: 5. See also ibid.: 147.

17 Ibid.: 148.

18 Carver, 1980: 8 and 6–10 on the range of other attacks that included killing or kidnapping military and police personnel, attacking military or police installations and attacking the railway system and even an oil refinery. The attack on the King David Hotel was politically counterproductive because of the 'collateral' deaths of dozens of ordinary civilians, including some Jewish civilians, but this apparent case of indiscriminate killing of civilians resulted from the failure of Irgun's telephone warning to evacuate the hotel (Begin, 1977: ch. 15).

19 O'Neill, 1990: 26.

20 Ibid.: 30 n. 18.

21 Ibid.: 45–7.
22 Gray, 2005: 212–13.
23 Jenkins, 1986: 783, 785–6.
24 Howard, 2002: 8.
25 However, Sambanis points out that such casualties criteria as the 1000-deaths-per-annum threshold can be problematic and not a very accurate indicator of the level of conflict. For example, they are not calculated on a per capita basis and therefore a small country's conflict may be excluded from a list of civil wars or insurgencies, as in the case of the Dhofar Rebellion in Oman and the Greco-Turkish war in Cyprus, even though proportionately the number of casualties was greater than in some listed conflicts that occurred in larger countries (Sambanis, 2004: 821–2).
26 Such a state is more of a 'juridical' than an 'empirical' entity (Jackson, 1990: 5). Reno's concept of the 'shadow state' is rather similar but in this case the growth of a neopatrimonial and clientelist shadow state has destroyed the bureaucratic aspects of statehood and created an empirically 'collapsed state' that comes close to being the opposite of classical definitions of the state (Reno, 2000: 46–7).
27 Therefore semistates are often the products of civil wars, which in recent years have come to be re-conceptualised as not necessarily 'engendering complete economic breakdown and anarchy' but instead having the 'potential' to create an alternative territorial form of power, protection and profit (Ballentine and Sherman, 2003: 2).
28 Mao, 1966a [1936]: 139.
29 Silber and Little, 1996: 214. My emphasis.
30 Ibid.: 171.
31 Ibid.: 244, 245.
32 Ahmad, 1993: 175, 176, 179. Hale describes the main source of rightist/leftist killings as the 'armed clashes' between such 'leftist organisations' as the Federation of Revolutionary Youth and 'the armed partisans of the extreme right', such as the Grey Wolves and the Association of Idealist Youth (Hale, 1994: 225).
33 Hamill, 1986: 163.
34 Waltz, 1979: 96–7. Waltz's neo-realist model of international politics is explicitly analogous to economists' model of firms' behaviour in the marketplace, which includes the possibility of a 'price war' breaking out between firms 'as part of a renewed struggle for markets' (Waltz, 1979: 88–98, 106). An actual, military war between states is presumably analogous to a price war between firms, but insurgent stateless war is analogous to what economists term 'countervailing power'. According to Galbraith's theory of countervailing power, a firm which 'enjoys a measure of monopoly power' in selling to its customers may be 'held in check by the countervailing power of those who are subject to it' if these customers 'develop the power with which they can defend themselves against exploitation' (Galbraith, 1963 [1952]: 125). The political analogy of this countervailing power involves the state's monopoly of force being held in check by the countervailing power of citizens, which enables them to defend themselves against exploitation by the state. Galbraith notes that customers may 'have a variety of weapons at their disposal to use against the market power of their suppliers' (ibid.: 134). And in the analogous political

relationship some citizens may resort to actual, military weapons and to an extreme kind of countervailing power – insurgent warfare against the state.

35 Clausewitz, 1976 [1832]: 480, 481.
36 Currey, 1997: 178.
37 Lautenschlager, 1986–7: 131.
38 Buckley, 1999: 2. My emphases.
39 Huntington, 1962: xviii.
40 Ibid.: xix. My emphases.
41 Thomas, 1929: 5 on comparison with pirates, 352 on effectiveness of the 'undersea raiders' and 3 on comparison with a celebrated German surface raider or 'corsair'.
42 Lautenschlager, 1986–7: 94–8, 140.
43 Ibid.: 105.
44 Ibid.: 101.
45 Huntington, 1962: xx.

Chapter 3

1 Arreguín-Toft, 2001: 105.
2 Clausewitz, 1976 [1832]: 481.
3 De Lee, 1988: 142–9; Dreyer, 1995: 249–53.
4 Westad, 2003: 31, 35, 112, 119–23, 174–5, 196.
5 Brooker, 2000: 95.
6 Even those who downplay the role of foreign military assistance in the Vietnamese case acknowledge that the 1975 offensive relied upon very conventional attacks carried out by regular units of North Vietnam's army (Gates, 1990: 342). In the Cambodian case, an army of tens of thousands of North Vietnamese regulars conquered large sections of the country during 1970–1 and enabled the Khmer Rouge guerrilla band to prepare for the semistate warfare that brought its final victory in 1975 (Short, 2004: 203–5, 216, 221).
7 Wright, 1991: 17–19.
8 Sweig, 2002: 14, 16, 21. The guerrillas were completely dependent upon the urban movement 'for everything from medicines, weapons, ammunition, food, equipment, clothing, money, and domestic and international publicity' (ibid.: 14).
9 Ibid.: 17, 120. See ibid.: 136–7, 152–3 on, however, the failure of the April 1958 uprising and the shift to a rural-oriented strategy.
10 Debray, 1968: 20–1, 63.
11 Chavarria, 1982: 34–7; Gorman, 1982: 118 on acquiring attributes of a regular army, such as the use of artillery and of staff officers.
12 Goodwin, 2001: 204–5.
13 Ibid.: 247.
14 Giri, 2008: 284, 286.
15 Ibid.: 282.
16 Wrong, 2005: 168, 180–3. See ibid.: 232, 280–1 and Tareke, 2002: 473–4 on the 1960s–70s.
17 Tareke, 2002: 477–80.

18 Tareke, 2004: 247–53.
19 Wrong, 2005: 354–5.
20 Stokke, 2006: 1038 n. 7.
21 McPherson, 1996: 117.
22 Ibid.: 156–7, 117. On the other hand, Lee's military victories had also prolonged a war that might otherwise have ended in 1862, when the Confederacy was 'on the verge of collapse' (ibid.: 158).
23 Joshi, 1996: 26. However, the federal authorities began to crack down on LTTE activities in Tamil Nadu in 1990, and 'the screws were tightened' further in 1991 after the LTTE's assassination of the former Prime Minister of India, Rajiv Gandhi (ibid.: 28, 29).
24 Stokke, 2006: 1022.
25 Ibid.: 1027–8, 1034.
26 Zartman, 1993: 24–5, 26.
27 Ibid.: 26.
28 Mao, 1966c [1938]: 214, see also 210–14.
29 Asprey, 1975: 515, 525.
30 Ibid.: 519, 526.
31 Bennett, 1988: 73–80.
32 Asprey, 1975: 501, 503, 505.
33 Quoted by Asprey, 1975: 472.
34 However, the one-third of Resistance insurgents under communist leadership rejected this strategy and instead 'sought to damage the Germans wherever and whenever they could be attacked, at whatever cost in reprisals, and whatever the strategic futility of their actions' (Hastings, 1993: 42).
35 Windrow, 1997: 3–4, 12.
36 Heggoy, 1972: 173.
37 Windrow, 1997: 11.
38 Windrow, 1998: 11. Of these about 48,000 were African colonial troops, 53,000 Vietnamese under French command, and 55,000 Vietnamese local auxiliaries.
39 Fall, 1967: 113–14, 116–17.
40 Windrow, 1998: 9–10.
41 Fall, 1967: 224.
42 Laqueur, 1977: 276. My emphases.
43 Roy, 1991: 23, 55 on Massoud and his practice of shifting to semistate warfare to take and hold enemy bases and towns. See ibid.: 56–7 on Hekmatyar's party, including the assessment that it was the only Islamist party 'to retain its demand for an Islamist revolution' rather than reverting during the course of the war to 'a more conservative fundamentalism' (ibid.: 57). Coll described Massoud as 'an attentive student of Mao and other successful leaders' and as 'a serious, deeply read student of Mao Zedong, Che Guevara and French revolutionary strategist Regis Debray'; Massoud also clearly established a semistate in the Panjshir valley, complete with police and taxes (Coll, 2005: 4, 116, 118).
44 Roy, 1991: 32. He estimates the mujaheddin's 'armed fighters' at 150,000–200,000 and the Soviet troops at 115,000 (ibid.: 17, 20, 32).
45 Quoted by Gott, 1970: 363.
46 Ibid.: 361; Laqueur, 1977: 316.

47 Debray, 1968: 61, 62.
48 Wright, 1991: 67.
49 Ibid.: 67, 94.
50 Debray, 1968: 53, 107–8.
51 Taber, 1970 [1965]: 37, 46–7.
52 Ibid.: 46–7, 91, 98.
53 Laqueur, 1977: 393, 406.
54 Ibid.: 393, 409.
55 Miller, 1980: 147 on membership figures. The National Liberation Movement adopted the informal name Tupamaro in 1965. It had first been used by followers of an Inca chief who had led a revolt against Spanish rule and later was used by a hero of the 1818–20 wars of Uruguayan independence (ibid.: 139).
56 Porzecanski, 1973: 17, 24.
57 Moss, 1972: 223, 238.
58 Guillen, 1982 [1973]: 317, 318–19, 321.
59 Morrison-Taw and Hoffman, 1994: 14.
60 Kepel, 2002: 262–3, 289–91. The Egyptian case occurred in 1992 and the Algerian cases in 1993–94.
61 Hamill, 1986: 68, and 104 on the No–Go areas as rival mini-states.
62 Ibid.: 112–17.
63 Bennett, 1988 : 69.
64 Anderson, D. L., 2005: 67; Duiker, 1981: 27.
65 Hashim, 2006: 43, 45.
66 Ibid.: 45, 46.
67 See on ideology Kramer, 1990 and Abu-Amr, 1994: 63–6, 80–2.
68 Mao, 1966c [1938]: 212.
69 Ibid.: 223, 258.
70 Quoted by Fall, 1967: 112.
71 Ibid.
72 Schelling, 1966: See 6 on Jewish 'terrorism' having an influence on British decisions, 29 on hurting people because they exert some influence on 'whether the war is continued', and 8–9 on Churchill's decisions not being influenced by 'a few casualties' but might have been influenced by mass panic if the German bombing of London had in fact created such panic. See also the quoted remarks by Sherman and by Fuller (ibid.: 15, 15 n. 7).
73 Schelling, 1966: 5.
74 See Wright, 1991: 112–13 on this case.
75 Ibid.: 113.
76 Hamill, 1986: 84; Coogan, 2002: 394.
77 Mack, 1975: 177, 179.
78 Clausewitz, 1976 [1832]: 91.
79 Ibid.: 92.
80 Such non-vital wars are also very different from the comparatively symmetric, 'zero-sum struggle for ascendancy' that is found in post-colonial wars between indigenous insurgents and a settler regime, such as in 1960s–70s Rhodesia/Zimbabwe, which are 'essentially a civil war in which the settler class has a survival stake in the outcome' (Mack, 1975: 191).

81 Mack, 1975: 182, 187, 189. He pointed to the opportunity costs of economic resources spent on the war and therefore unable to be allocated to other projects, and he noted that if the war escalates to such an extent as did the Algerian and Vietnamese wars, then tax increases may well be needed and inflation is a likely by-product of the war (ibid.: 185–6).
82 Ibid.: 181.
83 Brooker, 2005.
84 Mack, 1975: 192–3. He noted that the Portuguese empire was the last to be decolonised because of not only the presence of Portuguese settlers in the colonial territories but also the fact that retaining the colonies was important to a country that was too economically backward to win future economic benefits from these territories (through investment and trade in competition with other Western countries) once they became politically independent (ibid.: 190–2).
85 Ibid.: 189. He was describing the Portuguese dictatorship of the early 1970s and the controversies over the costs and benefits of continuing with its counterinsurgency wars in the colonies.
86 Anderson, D. L., 2005: 69.
87 Ibid.: 71.
88 Clausewitz, 1976 [1832]: 91.
89 Merom, 2003: 225.
90 Ibid.: chs 13 and 14.
91 Hashim, 2006: 178, 193, 195, 198. Although there was puzzlement in the US about the insurgents' goals 'beyond expelling the Americans', the insurgent wide-ranging attacks had such simple goals as to 'ensure that the American project of reconstructing Iraq fails' (Benjamin and Simon, 2005: 34).
92 Hashim, 2006: 47. See also Benjamin and Simon, 2005: 45.
93 Ricks, 2009: 46–7, 142, 200. Resettling the population is a counterinsurgency strategy that dates back to the Spanish colonial state's 1890s 'reconcentration' of the Cuban peasantry in the towns in an attempt to neutralise the Cuban nationalist insurgents (Porch, 2006: 151–2).
94 Ricks, 2009: 165–70, 202–5, 215–16, 219–21, 264–5. See Galula, 1964: ch. 7 on this classic counterinsurgency strategy. In comparison, resettling the population is not only a more drastic but also a desperate strategic remedy: 'Resettlement clearly is a last-resort measure, born out of the counter-insurgent's weakness' (ibid.: 112).
95 Gelpi, Feaver and Reifler, 2005/6: 44–5.
96 Ibid.: 41. The benefits were identified/chosen by survey respondents and in that order of preference.
97 Luttwak, 1995: 115.
98 Mueller, 2005: 45. The definition of 'disenchantment' with a war was more than half of opinion-poll respondents agreeing that the war was a mistake.
99 Merom, 2003: 19.
100 Ibid.: 21.
101 Mueller, 2005: 47.
102 Mack, 1975: 195.
103 See Arreguín-Toft, 2001: 119, 121–3 on importance of time and of public expectations of a quick victory.

104 Smith, 1995: 165.
105 Galula, 1964: 9.

Chapter 4

1 Jenkins, 1986: 776.
2 Grayling, 2007: 293.
3 Schelling, 1966: 16.
4 Ibid.
5 Overy, 2000: 89–90.
6 Robinson, 1980: 49, 56, 95–6.
7 Ibid.: 2–3, 50, 348–9. Most of the raids were on London and carried out in 1915–16.
8 Van Creveld, 2000: 167.
9 Quoted by Grayling, 2007: 131, 133.
10 In fact the British government had become committed to strategic bombing as a means of 'terrorising a European enemy from the air' and providing a 'terror bombing capability' that might deter an aggressor (Overy, 1980: 13). In contrast, the German air force was sceptical about 'whether strategic bombing by itself could achieve a decisive result by destroying industry or terrorising civilians' and its assessment of strategic bombing during the Spanish Civil War, notably the notorious bombing of Guernica, was quite negative: 'Terror bombing had had, for the most part, a counterproductive effect' (Murray, 1977: 110).
11 Overy, 1980: 24.
12 Overy, 2000: 88, 98.
13 Ibid.: 108–9 on bombing-related deaths in the Battle of Britain and the Blitz.
14 Grayling, 2007: 47, 50.
15 Overy, 1980: 121.
16 Schelling, 1966: 17, 30.
17 Grayling, 2007: 77–8.
18 Ibid.: 142.
19 Schelling, 1966: 17, 18.
20 Ibid.: 19–21.
21 Ibid.: 6.
22 Ibid.: 27.
23 Ibid.: 21, 25.
24 Pike, 1966: 252.
25 Al Qaeda's 9/11 attack on the US homeland is quite a different sort of exception because it was actually part of a provocation strategy, not a pressure strategy, and therefore is discussed in the chapter's section on the typically terrorist strategies of propaganda and provocation.
26 Crenshaw Hutchinson, 1978: 95–6.
27 Ibid.: 97, 100.
28 Bowden, 1976: 428. My emphasis.
29 Smith, 1995: 63.
30 Ibid.

31 Bowden, 1976: 432.
32 Ibid.: 436.
33 Smith, 1995: 124.
34 Hamill, 1986: 175, 176.
35 Smith, 1995: 128.
36 Ibid.: 195.
37 Ibid.
38 Hopgood, 2005: 56, 59.
39 Pape, 2003: 343, 358–9. The attacks included a) many combat or clearly military operations, b) many assassinations of senior officers, politicians and state officials, c) several attacks on economic targets.
40 Ibid.: 352. His other two clear-cut cases were Hezbollah's 1983 expulsion of US and French peacekeeping forces from Lebanon and the 1985 expulsion of Israel from much of southern Lebanon.
41 Abrahms, 2006: 47. These groups had been designated in 2001, and therefore excluded the, by then, pacified Provisional IRA, but most of the groups had been operating since before 1990 and in the majority of cases since the 1960s (ibid.: 47).
42 Ibid.: 55 and 50 Table 1. Apart from the LTTE, the only successful group among the 28 was Hezbollah, which was categorised as 'mainly military' in targeting and having 'total success' in the two cases of expelling 'peacekeepers' and 'Israel'.
43 Ibid.: 47 and 47 n. 25.
44 Crenshaw, 1990a: 11.
45 Cottrell, 2006: 35.
46 Sluka, 1997: 42.
47 Cottrell, 2006: 80.
48 Crenshaw, 1990a: 17.
49 Ibid.: 18.
50 Wallas, 1924 [1908]: 175.
51 Meijer, 2005: 283–4.
52 Ibid.: 285.
53 Moss, 1972: 223.
54 Gott, 1970: 129.
55 Laqueur, 1987 [1977]: 48.
56 Townshend, 2002: 59–60. He describes this era as the first age of terrorism, and seems to date the second age of terrorism as beginning in the late 1960s (ibid.: 54, 66).
57 Camus, 1971 [1951]: 133–8.
58 See Hoffman, 1998: 132 on Jenkins.
59 Dershowitz, 2003: 31.
60 Cook, 2005: 152.
61 Schelling, 1966: 3.
62 Horne, 1987: 186.
63 Ibid.: 206–7.
64 Laqueur, 1987 [1977]: 74.
65 Taber, 1970: 92, 97.
66 Merom, 2003: 22–3.
67 Laffan, 2004: 58–9. The insurgency that followed the successful secession from Britain was a rare event in the history of secessions and of anti-colonial strug-

gles. Kissane identified only 14 other cases of civil war occurring immediately after a secession or decolonisation in the period 1922–93, of which only seven were cases of 'civil war resulting from internal divisions within the nationalist movement', and in only one case, Algeria, was the political settlement or 'treaty' at issue (Kissane, 2005: 18, 19).

68 Marighella, 1971: 40.

69 The counterinsurgency campaign of 1969–70 included the widespread use of torture and the sending of military columns into the big cities to locate and destroy the insurgent bands in a series of one-sided street battles (Moss, 1971: 14).

70 Moss, 1972: 169.

71 Gott, 1970: 131.

72 Waldmann, 1986: 277.

73 Ibid.: 272–3; Goodwin, 2001: 186–7.

74 Wright, 1991: 183–4.

75 Goodwin, 2001: 223.

76 Ibid.: 237.

77 Ashworth, 1991: 102.

78 Byman, 1998: 155.

79 Gurr, 1990: 94, 95. In fact he suggested as a general principle that in democratic societies any terrorist violence offends most members of society and undermines the terrorists' support (ibid.: 102).

80 Kalyvas and Sanchez-Cuenca, 2005: 218–19. In addition to the outrage caused by attacks on civilians, suicide attacks may be contrary to some sectors of society's religious beliefs and/or moral attitudes. For example, the IRA found that the hunger strikes of the early 1980s increased their public support but that these self-inflicted deaths were opposed by the Catholic Church because of its doctrinal prohibition of suicide – and the opposition would have been all the stronger if the suicide had occurred as part of an act of violence (see Coogan, 2002: chs 28 and 29).

81 Kalyvas and Sanchez-Cuenca, 2005: 219.

82 Lawrence, 2005: 119.

83 Kepel, 2002: 2.

84 Ibid.: 4. This assessment of 9/11 is independently confirmed by Holmes, who infers the 'communicative strategy' as being 'an act of counter-propaganda' and an indirect strategy of 'provoking the United States into lashing out indiscriminately against Muslims around the world, thereby politicizing and radicalizing groups that had hitherto remained passive' (Holmes, 2005: 161, 162).

85 Kepel, 2002: 18.

86 Kilcullen, 2009: 34. There does not seem to be a planned provocation strategy involved in what he refers to as the accidental-guerrilla syndrome: 1) Al Qaeda establishes a safe haven 'in a remote, ungoverned or conflict-affected region', 2) from this safe haven Al Qaeda spreads its ideology and violence to other regions of the country or world, 3) forces from outside the region and/or the country 'intervene to deal with the AQ threat and disrupt the safe haven' and 4) the local population 'reacts negatively, rejecting outside intervention and allying with AQ' – and becoming 'accidental' guerrillas fighting against the intervening forces (ibid.: 35 Fig. 1.1 and 37 on external intervention).

87 Laqueur, 1977: 262, 271.
88 Schultz, 1978. See also Pike, 1966: 250 on the discriminating and selective use of terror.
89 Schultz, 1978: 81, 84.
90 Ibid.: 82, 83, 88.
91 Heggoy, 1972: 174–5.

Chapter 5

1 Hills, 2004: 147.
2 Nagl, 2005: xiii.
3 Brooker, 2000: chs 3 and 4. It is true that some insurgents have apparently been intent on democratising a dictatorship or replacing foreign rule with local democracy but so have many military coups that have overthrown a civilian or military dictator.
4 Criss, 1995: 20; Kramer, 2004/05: 8.
5 Guevara, 1961: 16.
6 Hobsbawm, 1973: 164. He was referring to the 'noble' or Robin Hood type of social bandit.
7 Laqueur, 1977: 96.
8 Kalyvas, 2001: 105.
9 Hobsbawm, 1972: 21, 81, 99.
10 Ibid.: 17, 18, 19, 24, 88. Their social rebellion cannot go beyond a 'programme' of the 'defence or restoration of the traditional order', such as of the 'good old king' in contrast to his usurpers, evil councillors or local authorities, and thus it has been easier to integrate bandits into national-liberation movements than into revolutionary movements (ibid.: 26–7, 104).
11 Ibid.: 58.
12 Benjamin and Simon, 2005: 61–2.
13 Hobsbawm, 1972: 46, 56.
14 Wright, 1991: 105, 108; Labrousse, 1973: 36.
15 Labrousse, 1973: 88.
16 Wright, 1991: 113.
17 Hobsbawm, 1972: 125.
18 Hopgood, 2005: 55.
19 Hobsbawm, 1972: 21, 24.
20 Hess, 1973: 162. He also supplies a table that juxtaposes and contrasts the main characteristics of the 19th century rural, protection-system mafia with those of the 20th century urban, gangster mafia (ibid.: 163).
21 Coogan, 2002: 380.
22 Gann, 1971: 85.
23 Laqueur, 1987 [1977]: 109; Dickie, 2004: xiv–xv.
24 Laqueur, 1987 [1977]: 109.
25 Dickie, 2004: 222.
26 Wright, 1991: 112.
27 Burke, 2004: 13.
28 Woodiwiss, 2003: 21–2.

29 Ibid.: 23.
30 Dickie, 2004: 195–6, 204–5.
31 Ibid.: 207, 214.
32 Sageman, 2004: 171.
33 Kilcullen, 2009: 223. He also points out that the insurgency in southern Thailand is one of the most intense in the world, second only to the Iraqi and Afghan, with civilians accounting for more than 70 per cent of the more than 2,500 killed in 2004–7 in a region with a population of only 1.8 million (ibid.: 210, 212).
34 Sageman, 2004: 171.
35 9/11 Commission, 2005: 171.
36 Sageman, 2004: 48.
37 Hess, 1973: viii.
38 Ibid.: 76; Gambetta, 1993: 122–5.
39 Coogan, 2002: 465 and 466 on specialised cells. For a different perspective see Hamill, 1986: 235.
40 Gunaratna, 2002: 76–84 on Al Qaeda and Sageman, 2004: 140 on Jemaah Islamiyah. Sageman points out that Al Qaeda did not necessarily practice what they preached about the cell structure (ibid.: 167).
41 Hashim, 2006: 166, and 161–9 on the functionally specialised cells.
42 Puzo, 1998 [1969]: 56.
43 Hamill, 1986: 290.
44 Klerks, 2003: 101–2.
45 Ibid.: 102 and see 112 n. 9 on the Internet comparison.
46 Sageman, 2004: 137–8, 140 on small-world, and 170 on seeming compartmentalisation, such as using different logistical support for their respective operations. See ibid.: 46 for definitions of these two small-world clusters: the Maghreb Arab and the Core Arab. But these apparently regional-origins definitions are confused by the fact that it is the origins of the *link* between a new member and the jihad that is crucial. For example, the Hamburg clique belonged to the Core Arab cluster, despite half the clique being originally from the Maghreb, because the original link had been a member of the Core Arab cluster (ibid.: 170). Furthermore, Sageman describes these two networks as each having an individual person as their hub: Khalid Sheik Mohammed as the hub of the Core Arabs and Abu Zubayadah as the hub of the Maghreb Arabs (ibid.: 137). Khalid Sheik Mohammed became an Al Qaeda ally or associate only in 1997–8 and thereafter became a formal member, while Abu Zubayadah was a 'longtime ally of Bin Laden' who apparently never became more than 'an associate' of Al Qaeda (9/11 Commission, 2005: 149, 154, 174, 435).
47 Burke, 2004: 327 n. 8.
48 Sageman, 2004: 140–1. See also ibid.: 142, 173 on the problems facing the relatively hierarchical JI cluster when it was 'decapitated' by the 2002–2003 arrests of its leadership.
49 Byman, 1998: 165.
50 Sageman, 2004: 141.
51 Burke, 2004: 264.
52 Ibid.: 267.
53 Ibid.: 265.

54 Benjamin and Simon, 2005: 27, 31.
55 Ibid.: 6–14. Burke, too, describes the Madrid attack as having 'little or no direction' from the AQ central leadership and being 'as autonomous an attack as is possible' (Burke, 2004: 272).
56 House of Commons Report, 2006.
57 *The Economist*, 19/9/2006: 34.
58 Bovenkerk and Chakra, 2007: 34–5.
59 Klerks, 2003: 101, 110.
60 Meijer, 2005: 298.
61 Burke, 2004: 270; Hashim, 2006: 143–4.
62 Hashim, 2006: 153, 155–6, 157.
63 Ibid.: 154, 155.
64 Ibid.: 187.
65 Ricks, 2009: 36. Apparently, its most important competitive advantage was financial, with other bands running short of money and AQI having ample funds to win the support of young men looking for money as well as willing to operate as insurgents (ibid.).
66 Ibid.: 202–9, 264–5.
67 Adams, 1988. See ibid.: 210–12 on IRA (and UDA) protection rackets.
68 Moss, 1972: 223.
69 Marighella, 1971: 31.
70 Weinstein, 2007: 291.
71 Ibid.: 291, 293.
72 Hodes and Sedra, 2007: 39.
73 Ibid.
74 Weinstein, 2007: 291.
75 Laqueur, 2001: 213.
76 Ballentine, 2003: 273.
77 Bovenkerk and Chakra, 2007: 30, 39.
78 Ibid.: 37.
79 Laqueur, 2001: 220.
80 Benjamin and Simon, 2005: 9–13.
81 O'Neill, 1990: 17–21.
82 O'Neill, 2005: 28. Another new type that he added is the 'apocalyptic–utopian' that includes such fringe groups as Aum Shinrikyo (ibid.: 23–4).
83 Ibid.: 28.
84 Compagnon, 1998: 85–6.
85 Boas and Dunn, 2007b: 34–5.
86 Brooker, 2000: chs 3 and 4.
87 Finer, 1976 [1962]: ch. 4.
88 Goodwin, 2001: 219. He indicated that the criteria for categorising an insurgency as persistent should include not only lasting for at least a decade but also fielding at least 1,000 armed combatants at the peak of the insurgency (ibid.: 220).
89 Ibid.: 220, 221.
90 Finer, 1976: 79–80; Brooker, 2000: 96–7.
91 Finer, 1976: 26–7; Brooker, 2000: 87–8.
92 Crenshaw, 1990a: 13.
93 Ibid.: 14.

94 Ibid.: 15. The availability of dynamite does indeed appear to have affected the outlook or insurgent calculus of anarchist revolutionary insurgents, as apparently in 1892 alone there were more than a thousand dynamite bombings in Europe (Camus, 1971 [1951]: 134).
95 Crenshaw, 1990a: 16.
96 Waterman, 1981: 587.
97 Ibid.: 558, 570.
98 Finer, 1976: 27; Brooker, 2000: 87–8.
99 Murray and Millet, 1992: 1.

Chapter 6

1 Petersen, 2001: 9.
2 Ibid.: 9, 13.
3 Sageman, 2004: 155.
4 Weinberg in Pedahzur, 2006: 119.
5 Petersen, 2001: 13, 14.
6 Ibid.: 11, 77, 78.
7 Other things being equal, the results of the assessment will be the same for individuals from an urban culture as for those from a rural culture, for individuals from a middle-class culture as for those from a working-class culture, for individuals from a secular culture as for those from a religious culture, and for individuals from one ethnic/tribal culture as for those from a different ethnic/tribal culture or from a non-ethnic or non-tribal culture.
8 Kilcullen, 2009: 77.
9 Shultz and Dew, 2006.
10 Ibid.: 27–8.
11 Kalyvas, 2001: 104, 106.
12 Brooker, 2000: 97.
13 Cater, 2003: 28, 29.
14 Collier, 2000: 100, and 99 on co-ordination and time-consistency problems.
15 Olson, 1971 [1965]: 51.
16 Recent insurgencies in Iraq and Afghanistan have taken military remuneration a step further or backwards by pioneering the use of bonuses or 'piece rates' for such dangerous tasks as planting roadside bombs (Ricks, 2009: 195; Kilcullen, 2009: 85).
17 See Taylor, 1988: 66 on running the risk of tautology when moving away from the essential features of what he calls the 'thin' theory of rationality and collective action, which in this case means moving away from its limited and materialistic view of what constitutes selective incentives.
18 Olson, 1971 [1965]: 61.
19 Ibid.: 62, 63.
20 Taylor, 1988: 66–7, 84.
21 Ibid.: 84 and see also 68.
22 Petersen, 2001: 71.
23 Olson, 1971 [1965]: 61–2 n. 17.
24 Kalyvas, 2001: 108.
25 Weinberg in Pedahzur, 2006: 119.

26 Kramer, 2004/05: 59.
27 O'Neill, 1990: 65.
28 Apparently, Pashtun tribes consider revenge (*badal*) to be part of the behaviour expected of 'a true Pashtun' (Kilcullen, 2009: 75).
29 Olson, 1971 [1965]: 61 n. 17. Although he argued that it is not possible to prove whether an individual's action was motivated by morality or by some other incentives, this is also true of other selective incentives – where circumstantial evidence is used to infer that this incentive was the 'motivator'.
30 Popkin, 1979: 254.
31 See Muller and Opp, 1986 for a move in this direction that included such non-material selective incentives as 'expected benefits of conforming to behavioral expectations of important others, and expected social affiliation rewards' (ibid.: 485).
32 Calhoun, 1991: 69.
33 Ibid.: 68, 69.
34 Calder, 1999: 121.
35 Calhoun, 1991: 61–5.
36 Offer, 1995: 229.
37 Burke, 2004: 258.
38 Weinstein, 2007: 120, 121. See 116–25 for an overall account of the SL recruiting experience.
39 Ibid.: 125.
40 Sageman, 2004: 153.
41 Ibid.: 144, 153.
42 Ibid.: 172.
43 Benjamin and Simon, 2005: 59, 61.
44 Sageman, 2004: 157.
45 Ibid.: 135.
46 Popkin, 1979: 240.
47 Sageman, 2004: 91–2, 121.
48 Ibid.: 112, 113. He sometimes uses a sample of as many as 165 but in this case the sample was reduced to only a 100 by the need for adequate background information (ibid.: 92, 115).
49 Ibid.: 101, 108, 115, 157 and see 152–3 on clique networks, which are local dense networks with face-to-face and long-term bonds.
50 Ibid.: see 115, 120–3, 135, 142–5, 168–9 on the three stages, and see 92–3, 143, 145 on geographical scattering of jihadist mosques from the West to Southeast Asia.
51 Ibid.: 156.
52 Ibid.: 123; Benjamin and Simon, 2005: 39–40 on individuals and 58 describing specialised recruiters in Europe as 'travel agents'; Hashim, 2006: 145, 147, 149–51. But some Iraqi-bound jihadists from Saudi Arabia had joined a small group of religious extremists that might be a bunch-of-guys clique or were heeding the jihadist call of clerics and activists (ibid.: 148, 150).
53 Clark, 1984: 156–7.
54 Ibid.: 157. The reasons varied from wanting to complete their military service or university education to wanting to experiment with other organisations to see if these would be more effective than ETA as a way of pursuing Basque nationalism.

55 Benjamin and Simon, 2005: 8–9.
56 Begin, 1977: 46.
57 Ibid.: 66.
58 Sageman, 2004: 155.
59 Wood, 2003: 205 and see also 18, 235.
60 McCauley, 2001: 356.
61 Elster, 2005: 242, 243, 256.
62 O'Neill, 1990: 77.
63 Sageman, 2004: 93.
64 Elster, 2005: 241–2, 245.
65 Hopgood, 2005: 65.
66 Pape, 2003: 343.
67 Merari, 1990: 204, 205.
68 Offer, 1995: 229.

Chapter 7

1 Kalyvas and Sanchez-Cuenca, 2005: 211.
2 Van Creveld, 2000: 90. See ibid.: 89 for his description of Prussia as an 'artificial country' that was 'held together as a work of art'.
3 Spiers, 1975: 7, 12.
4 Hashim, 2006: 189.
5 Pape, 2003: 346.
6 Ibid.
7 Hill, 2005: 9, 10, 12.
8 Glantz and House, 1995: 377 n. 3.
9 Hopgood, 2005: 54–5.
10 Kalyvas and Sanchez-Cuenca, 2005: 210, 226. The willingness to use this form of attack depends upon an implicit cost-benefit analysis that takes into account the likely counterproductive effects, constituency costs and technological costs of using suicide rather than conventional forms of attack.
11 Weinstein, 2007: 289.
12 Ibid.: 280.
13 Castaneda, 1997: 347.
14 Ballentine, 2003: 271.
15 Collier, 2000: 101.
16 Townshend, 2002: 68, 73. Although the Baader-Meinhof group (which later called itself the Red Army Faction) was indeed tiny and did not carry out many attacks, the Red Brigades were a confederation of groups that in total launched a very large number of attacks, increasing from 467 in 1975 to a peak of 2,725 in 1978 (ibid.: 69).
17 Hodes and Sedra, 2007: 104 n. 48, 57.
18 Petersen, 2001: 3; Cottrell, 2006: 80.
19 Begin, 1977: 61.
20 Ibid.: 52.
21 Hashim, 2006: 163–4.
22 Hastings, 1993: chs 4 and 5.

23 Sageman, 2004: 53–4.
24 Benjamin and Simon, 2005: 76, 77, 79.
25 Cohen and Gooch, 2006: 94.
26 Kilcullen, 2009: 53.
27 Sageman, 2004: 90.
28 Ibid.: 106. And the 9/11 operation succeeded only because these operatives were indeed 'able to overcome the myriad of obstacles they encountered' (ibid.: 167).
29 Fall, 1967: 366.
30 Ibid.: 366–7.
31 Hashim, 2006: 190–2.
32 Benjamin and Simon, 2005: 42–3 citing such experts as Cordesman and Davies.
33 Kramer, 2004/05: 28.
34 Cohen and Gooch, 2006: 175.
35 Coogan, 1992: 76–83, 128.
36 Townshend, 1979: 326, 327, 329.
37 Bew, 2004: 76.
38 Edgerton, 1989: 73. The Mau Mau case is also an example of the importance of training – or the lack of it. In mid-1953 the bands living in the forests were organised and trained by leaders who had often served in African units of the British army, but they had seldom been in combat units, let alone in fire fights, and so could not provide much useful combat training (ibid.: 82, 112–13).
39 Anderson, D., 2005: 195.
40 Edgerton, 1989: 73, 130.
41 Kuperman, 1999: 244. Similarly, the Chechen separatist insurgency of the early 2000s acquired Strela and Igla missiles, the equivalent of Stingers, which helped the insurgents to neutralise the Russian helicopters' potential for reducing the insurgent-favouring effects of mountainous and forested terrain (Kramer, 2004/05: 32, 38, 42). In mountainous southern Chechnya 'terrain has been a key advantage for guerrillas, enabling them to ambush Russian forces, conceal ammunition and weapons, and move almost unhindered between Chechnya and safe havens across the border in Georgia, Dagestan, and Ingushetia' (ibid.: 6). But the Chechens' ability to exploit this terrain opportunity fully was due to their having high-tech air-defence equipment – one of the strengths of their means/capability variable.
42 Kuperman, 1999: 236, 248.
43 Norton, 2000: 30–1.
44 Ibid.: 30.
45 Chaliand, 1982: 16. My emphasis.
46 O'Neill, 1990: 67.
47 Debray, 1968: 45 footnote.
48 Porzecanski, 1973: 18.
49 Ibid.: 11, 12. My emphases.
50 White, 2003: 100.
51 Ibid.: 139.
52 See Cohen and Gooch, 2006: 63, 67, 72, 84 on the failure of US Navy to learn from British experience in anti-submarine warfare, and 161–3 on the

failure to adapt in the case of British landings at Suvla Bay in the 1915 Gallipoli campaign.

53 Dewar, 1992: 54.

54 Shultz and Dew, 2006: 82. Few other insurgent leaders could match Aidid's intellectual capacity – his wide range of ability and experience. This former General in Somalia's defunct army had been trained in Italy and Russia, had served as ambassador to India, spoke English as well as Italian and Russian and had published three books (ibid.: 92).

55 Ibid.: 68.

56 Ibid.: 93–4, 97.

Chapter 8

1 See Hashim, 2006: 33, 184–5 on the growing role of ex-army personnel in providing trained manpower and training expertise, 188 for examples of their military skills in ambushes, 163 on the thousands of ammunition dumps and hundreds of thousands of tonnes of munitions stored in them, and 33, 40, 191–2 on the insurgents' adaptability at operational and tactical levels.

2 Guevara, 1961: 16.

3 Mao, 1966a [1936]: 79, 80. My emphases.

4 Ibid.: 80.

5 Keegan, 1993: 63.

6 Galula, 1964: 35.

7 Mao, 1966b [1938]: 173.

8 O'Neill, 1990: 54, 57. However, as Mao warned, without sufficient territory to manoeuvre, their areas of operation 'can be cordoned off, isolated, turned into free-fire zones and penetrated' (ibid.: 55). O'Neill also concluded that the climate factor, too, was significant but hampered both sides of the conflict and does not appear to favour either side (ibid.: 57–8).

9 Galula, 1964: 36.

10 Ibid.: 38.

11 Fearon and Laitin, 2003: 81, 85. As was noted earlier, their definition of civil war as intra-state conflicts that killed at least 1,000 people, with a yearly average of at least 100, would cover major insurgencies as well as semistate wars (ibid.: 76).

12 Sambanis, 2004: 844, 856.

13 O'Neill, 1990: 58–9, quoting Gurr, 1970: 263–4.

14 Ballentine, 2003: 267.

15 Laqueur, 1977: 317.

16 O'Sullivan and Miller, 1983: 136 on growing proportion of land surface, 137 on growing proportion of population.

17 Ibid.: 127.

18 Ashworth, 1991: 88.

19 Ibid.

20 Hills, 2004: 57, 68, 132, 204, 207.

21 Griffin, 2002: 225.

22 Ashworth, 1991: 97.

23 Antal and Gericke, 2003: 419, 420.
24 Dewar, 1992: 78, 92. A recent analysis of urban combat pointed out that 'U.S. forces do not possess an overwhelming high technology advantage in urban combat as they do in virtually all other combat environments' (Antal and Gericke, 2003: 430).
25 Antal and Gericke, 2003: 429.
26 Ashworth, 1991: 98, 99. However, he believes that fear of the mob was not in fact the chief motivation behind the new fashion in town-planning (ibid.: 99).
27 Ibid.: 105.
28 Debray quoted by Gott, 1970: 135. See also Callanan, 1969 on the Venezuelan urban insurgents and how they were militarily defeated.
29 Morrison-Taw and Hoffman, 1994: 11.
30 Ibid.: 12.
31 Ashworth, 1991: 105.
32 Dewar, 1992: 9.
33 O'Sullivan and Miller, 1983: 131; Dewar, 1992: 13.
34 Coaffee, 2004: 277.
35 Marcuse, 2004: 271.
36 Ibid.
37 Mao, 1966a, b, c and d [1936, 1938].
38 Wickham-Crowley, 1992: 299.
39 Goodwin, 2001: 212–13 on confirming Wickham-Crowley.
40 Goodwin, 2001: 303 on Honduras and 193, 208, 245 on democracies not producing major revolutionary insurgencies. See also ibid.: 300, 302 on confirming Guevara.
41 An obvious recent case is the Algerian FIS/AIS's 1997 unilateral ceasefire and negotiated truce with the army. This was 'an "honourable" intermediate solution between defeat and surrender' which allowed the movement 'to keep its men together in the interval before they could be integrated into the security forces' (Kepel, 2002: 274). In contrast, Algeria's other main revolutionary insurgent band, the GIA, had become increasingly violent in its terrorist-mode attacks until its mysterious disappearance in late 1997. In addition to making incongruous bombing attacks in France, it ratcheted up the amount of indiscriminate and brutal violence that it inflicted on an Algerian population that had supposedly betrayed its hopes (ibid.: 267, 272–3).
42 Guevara, 1961: 16.
43 Dion, 1996: 270.
44 Ibid.: 269, 270.
45 Shultz and Dew, 2006: 94.
46 Goodwin, 2001: 212. Thus when he identified the particularly vulnerable variety of dictatorship he included more than just the political characteristics of being a neopatrimonial type of dictatorship engaged in politically counterproductive repression; he also specified that the dictatorship had a 'weak' state which lacked military power and other 'infrastructural' aspects of a state's strength. See ibid.: 235, 250 on 'infrastructural' weak/strong state in terms of control and 225, 228, 250–1 on military power.
47 Galula, 1964: 31.

48 Callwell, 1990 [1896]: 29, 33, 42.
49 O'Neill, 1990: 53–69.
50 Galula, 1964: 5–7, 75, 95.
51 Draper, 1967: 30.
52 Thompson, 1966: 51.
53 Huntington, 1962: xx. My emphasis.
54 Nagl, 2005: xiii.
55 Galula, 1964: 32.
56 Joes, 1996: 134, 135.
57 Ozdag and Aydinli, 2003: 113 and 120 n. 45.
58 Porch, 2006: 208.
59 Kalyvas, 2006: on insurgent-dominated 211, 238; on state-controlled 211, 222, 223.
60 Galula, 1964: 86.
61 Trotsky, 1934: 225. Kalyvas, 2006: 18 acknowledges Trotsky's (and Tilly's) influence on his conception of civil war as 'contestation of sovereignty'.
62 Kalyvas, 2006: 88.
63 Ibid.: 132–8 on distribution of control; 211–12 on the variation of control across the five zones.
64 Ibid.: 213, 218–19. So it is not surprising that a similar typology to Kalyvas's was presented by Galula in his military counterinsurgency handbook (Galula, 1964: 55–6). In this typology, insurgents set up various types of areas: regular base areas that are safe from all but a major military campaign by the counterinsurgents (equivalent to zone 5); guerrilla base areas that are subject to more frequent but usually temporary penetrations by counterinsurgent forces (equivalent to zone 4); guerrilla areas, in which insurgents are 'constantly contending' with counterinsurgent forces and administration (equivalent to zones 2 and 3); and occupied areas that are 'under the counterinsurgent's political and military control' and 'where the insurgent works only underground' (equivalent to zone 1).
65 Chongyi and Goodman, 2000: 9, 10.
66 Kalyvas, 2006: 368, 370, 374–6.
67 Ibid.: 365. My emphases. See also ibid.: 383 for a different formulation of the definition.
68 O'Neill, 1990: 71–2.
69 Begin, 1977: 109. My emphases.
70 O'Neill, 1990: 71–2.
71 Short, 1975: 375–8.
72 Gunaratna, 2003: 221 n. 37.
73 9/11 Commission, 2005: 170.
74 Sageman, 2004: 94. The full-timers were most prevalent in what he terms the Central Staff of Al Qaeda and in the Core Arab network, over half of whom were engaged in full-time terrorist activities.
75 Clark, 1984: 158.
76 Petersen, 2001: 9.
77 Kilcullen, 2009: 84–5.
78 Gurr, 1990: 94, 95.
79 Sprinzak, 1990: 77.
80 Ibid.: 78.

81 Galula, 1964: 19.
82 Malaquias, 2007: 217.
83 Ibid.: 218.

Chapter 9

1 Gott, 1970: 363.
2 Chongyi and Goodman, 2000: 8.
3 Keating, 2000: 31, 32.
4 Boas and Dunn, 2007a: 3, 4.
5 Laqueur, 1987 [1977]: 167.
6 Ibid.: 164–5. He therefore made a devastating critique of the frustration-aggression psychological theory, originally developed in the 1930s, that violence resulted from the discontent caused by frustration – from the discrepancy between the level of social wants and the extent to which those wants were satisfied (ibid.: 158–9). In his view all that can be established about the likelihood of a small group mobilising themselves for a terrorist campaign is that such 'terrorism is more likely to occur in certain conditions than in others, and that in some conditions it cannot take root at all', namely those conditions to be found in effective dictatorships and in some small democracies with a peaceful political culture (ibid.: 172, 173).
7 Sageman, 2004: 95.
8 Ibid.: 119.
9 Gurr, 1970: 319, 357.
10 Ibid.: 359.
11 Ibid.: 358.
12 Ibid.: 319.
13 Galula, 1964: 18–19.
14 Ibid.: 23, see also 21–4.
15 Weinstein, 2007: 12. My emphases.
16 Taylor, 1988: 80–1. See also 73–7 on China and 77–80 on the 'rational peasant' in Vietnam and the party's efforts to win him over.
17 Popkin, 1988: 49–50, 61, 62.
18 Cheibub, 1995: 247–8.
19 Ibid.: 248.
20 Goodwin and Skocpol, 1989: 493.
21 Ibid.: 493, 494.
22 Kilcullen, 2009: 58.
23 Galula, 1964: 76, 128.
24 See Paget, 1967: 127–8, 134, 143 on popular support for the insurgents and ibid.: 145, 147 on failure of Malayan methods and of information-gathering from the population.
25 Galula, 1964: 77, 130.
26 Kalyvas, 2006: 102, 103.
27 Ibid.: 124, 128–9.
28 Scott, 1985: 29.
29 Tucker, 1995: 61–2.

30 Weinstein, 2007: 133–4. Technically speaking, peer monitoring shapes other individuals' (rational) 'choices about the level of effort they will exert' and thereby reduces 'shirking' (ibid.: 133).
31 Kalyvas, 2006: 102, 218 and 124, 128 on measures that change preferences. Although at times he seems confident that coercion can produce a non-shirking 'collaboration', clearly this use of unadulterated coercion is an inefficient means of eliciting active support (ibid.: 104). For using insurgent forces to coerce a support base is clearly an inefficient use of military resources, and in terms of effectiveness, too, it pays to mix coercion with large amounts of persuasion.
32 Weinstein, 2007: ch. 5. He depicted these political measures in the institutionalist terms of rulers constructing political institutions that provide them with a credible commitment 'to show restraint in the extraction of resources' from their subjects (ibid.: 169).
33 Kilcullen, 2009: 83.
34 Ibid.: 85, 86.
35 Benton, 2000: 202–3, 213–14.
36 Kilcullen, 2009: 78.
37 Ibid.: 82 Figure 2.3.
38 Weinstein, 2007: 151.
39 Wood, 2003: 2–3.
40 Ibid.: 17.
41 Ibid.
42 Ibid.: 233, 237.
43 Kalyvas, 2006: 190, 191.
44 Meijer, 2005: 282, 299.
45 Ibid.: 282.
46 Ibid.: 299.
47 Crenshaw Hutchinson, 1978: 43, 146.
48 Ibid.: 47–8.
49 Ibid.: 148. My emphases.
50 Ramakrishna, 2003: 167.
51 Crenshaw Hutchinson, 1978: 48, 148. It is true that some individuals and communities instead showed support, worked and fought for the French but this was for primarily non-ideological motives (Kalyvas, 2006: 47).
52 Thompson, 1966: 63.
53 Liberman, 1996: 20. As examples, he points to ideologically divided China fighting against the Japanese invaders and to Tito leading ethnically-divided Yugoslavians against the Nazi German invaders.
54 O'Neill, 1990: 61, 65.
55 Ibid.: 62.
56 Byman, 1998: 154.
57 For example see Stepan, 1998: 232–3.
58 Byman, 1998: 150, 154.
59 Toft, 2003: 19.
60 Ibid.: 21, 23.
61 Ibid.: 21, 22.
62 Putnam, 1993: 175. My emphases.
63 Ibid.: 172.

64 Kilcullen, 2009: 251 argues that Europe's immigrant Islamic communities are better viewed as 'a *target* of terrorist-sponsored subversion than a *source* of threat'. But they also seem to be a source of what he terms 'micro-havens' where 'the writ of government does not always run and where police and security services' situational awareness is low' (ibid.: 246). In fact these micro-havens may well be islands of O1, O2 or O3 control within which potential political support can be converted into actual support and significant support bases established.
65 McCauley, 2001: 344.
66 Calhoun, 1988: 172.
67 See Compagnon, 1998, on the complexities of Somali clans' (and intra-clan segments') political mobilisation. See Kalyvas, 2006: 373 n. 11 on clans among Iraqi Sunni and Shiites. 'Tribes were less significant in the cities and among Shiites', though, than in rural areas and among Sunnis (Ricks, 2009: 72).
68 Calhoun, 1988: 170, 174.
69 Waterman, 1981: 562–4.
70 Ibid.: 565.
71 Kuran argued that a person 'may choose to voice a demand for change even when the price of dissent is very high and the chances of success very low' if the 'private opposition is very intense and/or his need for integrity is quite strong' – in other words, where the grievance and/or moral selective incentive is unusually strong (1991: 18–19, 20).
72 Ibid.: 20, 22, 24. The latent bandwagon and the chain reaction are similar to the concept of an informational cascade that Lohmann applied to the street demonstrations that peacefully toppled the East German communist regime. In this case the population is being given informational cues or signals by the increasing size of the protest demonstrations but what is being revealed is the malignity of the regime rather than its weakness; the protesters are providing costly and credible signals of their negative experience of the regime (Lohmann, 1994: 49, 51–2). She argues that her model and Kuran's can be synthesised, with the former covering the (negative) nature of the regime and the latter covering the (declining) cost of participating in opposition (ibid.: 54–5).
73 Petersen, 2001: 268.
74 Ibid.: 245, 247. His model is also based on risk aversion (the bravery/cowardice factor) and on susceptibility to regime threat (the opportunity-cost fear of failure).
75 Kuran, 1991: 25. He recognised the role of revolutionary leaders in an earlier presentation of his theory (Kuran, 1989: 64–5).
76 Wright, 1991: 185.
77 Goodwin, 2001: 192.

Conclusion

1 Van Creveld, 1991: 197–8. In a later article he noted that he wrote this book in 1988–9 when 'the Soviet Union was still a functioning superpower' and the 'Cold War was still raging' (Van Creveld, 2002: 3).

2 Van Creveld, 2000: 213.
3 Murray, 1992: 89, 93.
4 Hirst, 2001: 99.
5 Murray and Knox, 2001: 13, Table 1.1.
6 Kellman, 2007: 54 quoting a famous 2000 article by Bill Joy in *Wired* magazine, issue 8.04.
7 Van Creveld, 2002: 9, 11, 12.
8 For example, new communications technology may have removed one of the main restrictions on the ability of the Internet to generate insurgent activity. Sageman has argued that 'intense lengthy face-to-face interactions' are required to create jihadists. 'Reading and sending messages about the jihad on the Internet may make these individuals receptive to its appeal, but direct involvement requires face-to-face interaction' (Sageman, 2004: 157). However, two-way visual communication through the Internet may provide sufficient face-to-face interaction to stimulate direct involvement.
9 Ellis, 2004: 123.
10 Kellman, 2007: 51–2.
11 Hirst, 2001: 90, 91.
12 See Boas and Dunn, 2007b: 19–20 for a recent discussion of how 9/11 and the 'war on terror' has led, in a manner reminiscent of the Cold War, to a new concern in the West about 'failed' states and other security threats coming from 'out of Africa'.
13 Simmel, G. 'The Sociology of Conflict, II', *American Journal of Sociology*, Vol. 8, March, 1904. Quoted by Waltz, 1979: 188.
14 Kalyvas, 2006: 139.
15 Waltz, 1979: 189.
16 Ibid.
17 Brooker, 2009: 27.
18 Huntington, 1991: 294.
19 Huntington, 1959: 346–50.
20 Brooker, 2005.

Bibliography

Abrahms, M. (2006) 'Why Terrorism Does Not Work', *International Security*, 31/2, 42–78.

Abu-Amr, Z. (1994) *Islamic Fundamentalism in the West Bank and Gaza* (Bloomington, Indiana).

Adams, J. (1988) *The Financing of Terror* (Sevenoaks, England).

Ahmad, F. (1993) *The Making of Modern Turkey* (London).

Anderson, D. (2005) *Histories of the Hanged: The Dirty War in Kenya and the End of Empire* (New York).

Anderson, D. L. (2005) *The Vietnam War* (New York).

Antal, J. and Gericke, B. (2003) 'Lessons Learned from City Fights', in J. Antal and B. Gericke (eds) *City Fights: Selected Histories of Urban Combat From World War II to Vietnam* (New York).

Arreguín-Toft, I. (2001) 'How the Weak Win Wars: A Theory of Asymmetric Conflict', *International Security*, 26/1, 93–127.

Ashworth, G. J. (1991) *War and the City* (London).

Asprey, R. B. (1975) *War in the Shadows: The Guerrilla in History* (New York).

Ballentine, K. (2003) 'Beyond Greed and Grievance: Reconsidering the Dynamics of Armed Conflict', in K. Ballentine and J. Sherman (eds) *The Political Economy of Armed Conflict: Beyond Greed and Grievance* (Boulder).

Ballentine, K. and Sherman, J. (2003) 'Introduction', in K. Ballentine and J. Sherman (eds) *The Political Economy of Armed Conflict: Beyond Greed and Grievance* (Boulder).

Beckett, I. F. W. (2001) *Modern Insurgencies and Counter-Insurgencies* (London and New York).

Begin, M. (1977 [1952]) *The Revolt* (Jerusalem).

Bell, D. A. (2007) *The First Total War: Napoleon's Europe and the Birth of Modern Warfare* (London).

Benjamin, D. and Simon, S. (2005) *The Next Attack: The Globalization of Jihad* (London).

Bennett, M. (1988) 'The German Experience', in I. F. W. Beckett (ed.) *The Roots of Counter-Insurgency* (London).

Benton, G. (2000) 'Comparative Perspectives: North and Central China in the Anti-Japanese Resistance', in F. Chongyi and D. Goodman (eds) *North China at War: The Social Ecology of Revolution 1937–1945* (Lanham, Maryland).

Berdal, M. and Malone, D. M. (2000) 'Introduction', in M. Berdal and D. M. Malone (eds) *Greed and Grievance: Economic Agendas in Civil Wars* (Boulder).

Bew, P. (2004) 'Moderate Nationalism, 1918–23: Perspectives on Politics and Revolution', in M. J. Bric and J. Coakley (eds) *From Political Violence to Negotiated Settlement: The Winding Path to Peace in Twentieth-Century Ireland* (Dublin).

Boas, M. and Dunn, K. C. (2007a) 'Introduction', in M. Boas and K. C. Dunn (eds) *African Guerrillas: Raging Against the Machine* (Boulder).

Boas, M. and Dunn, K. C. (2007b) 'African Guerrilla Politics: Raging Against the Machine?', in M. Boas and K. C. Dunn (eds) *African Guerrillas: Raging Against the Machine* (Boulder).

Bovenkerk, F. and Chakra, B. A. (2007) 'Terrorism and Organised Crime', in L. Holmes (ed.) *Terrorism, Organised Crime and Corruption* (Cheltenham, England).

Bowden, T. (1976) 'The IRA and the Changing Tactics of Terrorism', *Political Quarterly*, 47/4, 425–37.

Brooker, P. (1995) *Twentieth-Century Dictatorships* (Basingstoke).

Brooker, P. (2000) *Non-Democratic Regimes: Theory, Government and Politics* (Basingstoke).

Brooker, P. (2005) *Leadership in Democracy* (Basingstoke).

Brooker, P. (2009) *Non-Democratic Regimes*, Revised Edition (Basingstoke).

Buckley, J. (1999) *Air Power in the Age of Total War* (Bloomington and Indianapolis).

Burke, J. (2004) *Al-Qaeda: The True Story of Radical Islam* (London).

Burton, A. M. (1975) *Urban Terrorism: Theory, Practice and Response* (London).

Byman, D. (1998) 'The Logic of Ethnic Terrorism', *Studies in Conflict and Terrorism*, 21, 149–69.

Calder, A. (1999) 'The Diary of an Unknown Aviator', in A. Calder (ed.) *Wars* (London).

Calhoun, C. J. (1988) 'The Radicalism of Tradition and the Question of Class Struggle', in M. Taylor (ed.) *Rationality and Revolution* (Cambridge).

Calhoun, C. (1991) 'The Problem of Identity in Collective Action', in J. Huber (ed.) *Macro–Micro Linkages in Sociology* (Newbury Park, California).

Callanan, E. F. (1969) 'Terror in Venezuela', *Military Review*, 49/2, 49–56.

Callwell, C. E. (1990 [1896]) *Small Wars: A Tactical Textbook for Imperial Soldiers* (London).

Camus, A. (1971 [1951]) *The Rebel* (Harmondsworth).

Carver, M. (1980) *War Since 1945* (London).

Castaneda, J. (1997) *Companero: The Life and Death of Che Guevara* (London).

Cater, C. (2003) 'The Political Economy of Conflict and UN Intervention: Rethinking the Critical Cases of Africa', in K. Ballentine and J. Sherman (eds) *The Political Economy of Armed Conflict: Beyond Greed and Grievance* (Boulder).

Chaliand, G. (1982) 'Introduction', in G. Chaliand (ed.) *Guerrilla Strategies: A Historical Anthology From the Long March to Afghanistan* (Berkeley).

Chartrand, R. (2004) *Spanish Guerrillas in the Peninsular War 1808–14* (Oxford).

Chavarria, R. E. (1982) 'The Nicaraguan Insurrection: An Appraisal of its Originality', in T. W. Walker (ed.) *Nicaragua in Revolution* (New York).

Cheibub, J. A. (1995) 'Mobilizing and Sustaining Collective Action in the Mexican Revolution', *Politics and Society*, 23/2, 243–52.

Chongyi, F. and Goodman, D. (2000) 'Introduction: Explaining Revolution', in F. Chongyi and D. Goodman (eds) *North China at War: The Social Ecology of Revolution 1937–1945* (Lanham, Maryland).

Clapham, C. (1998) 'Introduction: Analysing African Insurgencies', in C. Clapham (ed.) *African Guerrillas* (Oxford).

Clark, R. P. (1984) *The Basque Insurgents: ETA, 1952–1980* (Madison).

Clausewitz von, C. (1976 [1832]) *On War* (Princeton).

Coaffee, J. (2004) 'Recasting the "Ring of Steel": Designing Out Terrorism in the City of London?', in S. Graham (ed.) *Cities, War and Terrorism* (Oxford).

Cohen, E. A. and Gooch, J. (2006) *Military Misfortunes: The Anatomy of Failure in War* (New York).
Coll, S. (2005) *Ghost Wars* (London).
Collier, P. (2000) 'Doing Well Out of War: An Economic Perspective', in M. Berdal and D. M. Malone (eds) *Greed and Grievance: Economic Agendas in Civil Wars* (Boulder).
Compagnon, D. (1998) 'Somali Armed Units: The Interplay of Political Entrepreneurship and Clan-based Factions', in C. Clapham (ed.) *African Guerrillas* (Oxford).
Constable, P. (2007) 'A Wake-Up Call in Afghanistan', *Journal of Democracy*, 18/2, 84–98.
Coogan, T. P. (1992) *The Man Who Made Ireland: The Life and Death of Michael Collins* (Niwot, Colorado).
Coogan, T. P. (2002) *The IRA* (New York).
Cook, D. (2005) *Understanding Jihad* (Berkeley).
Cottrell, P. (2006) *The Anglo-Irish War: The Troubles of 1913–1922* (Oxford).
Crenshaw Hutchinson, M. (1978) *Revolutionary Terrorism: The FLN in Algeria, 1954–1962* (Stanford).
Crenshaw, M. (1990a) 'The Logic of Terrorism: Terrorist Behavior as a Product of Strategic Choice', in W. Reich (ed.) *Origins of Terrorism* (Washington D. C.).
Crenshaw, M. (1990b) 'Questions to be Answered, Research to be Done, Knowledge to be Applied,' in W. Reich (ed.) *Origins of Terrorism* (Washington D.C.).
Crenshaw, M. (1995) *Terrorism in Context* (Philadelphia).
Criss, N. B. (1995) 'The Nature of PKK Terrorism in Turkey', *Studies in Conflict and Terrorism*, 18, 17–37.
Currey, C. B. (1997) *Victory at Any Cost: The Genius of Vietnam's Gen. Vo Nguyen Giap* (Washington).
Debray, R. (1968) *Revolution in the Revolution?* (Harmondsworth).
De Lee, N. (1988) 'The Far Eastern Experience', in I. F. W. Beckett (ed.) *The Roots of Counter-Insurgency* (London).
Dershowitz, A. M. (2003) *Why Terrorism Works* (Melbourne).
Dewar, M. (1992) *War in the Streets* (London).
Dickie, J. (2004) *Cosa Nostra: A History of the Sicilian Mafia* (London).
Dion, S. (1996) 'Why is Secession Difficult in Well-Established Democracies? Lessons from Quebec', *British Journal of Political Science*, 26/2, 269–83.
Draper, T. (1967) *Abuse of Power* (New York).
Dreyer, E. L. (1995) *China at War, 1901–1949* (London).
Duiker, W. J. (1981) *The Communist Road to Power in Vietnam* (Boulder).
Edgerton, R. B. (1989) *Mau Mau: An African Crucible* (New York).
Eldredge, N. (1995) *Dominion* (Berkeley).
Ellis, J. (2004) 'Terrorism in the Genomic Age', in R. Gunaratna (ed.) *The Changing Face of Terrorism* (Singapore).
Elster, J. (2005) 'Motivations and Beliefs in Suicide Missions', in D. Gambetta (ed.) *Making Sense of Suicide Missions* (Oxford).
Englebert, P. (2007) 'Whither the Separatist Motive?', in M. Boas and K. C. Dunn (eds) *African Guerrillas: Raging Against the Machine* (Boulder).
Fall, B. (1967) *The Two Vietnams: A Political and Military Analysis* (New York).
Fearon, J. D. and Laitin, D. D. (2003) 'Ethnicity, Insurgency, and Civil War', *American Political Science Review*, 97/1, 75–89.

Ferrill, A. (1985) *The Origins of War* (London).
Finer, S. E. (1976 [1962]) *The Man on Horseback: The Role of the Military in Politics*. Second, enlarged edition (Harmondsworth).
Finlan, A. (2004) *The Collapse of Yugoslavia 1991–1999* (Oxford).
Galbraith, J. K. (1963 [1952]) *American Capitalism: The Concept of Countervailing Power* (Harmondsworth).
Galula, D. (1964) *Counterinsurgency Warfare: Theory and Practice* (New York).
Gambetta, D. (1993) *The Sicilian Mafia: The Business of Private Protection* (Cambridge, Mass.).
Gann, L. H. (1971) *Guerrillas in History* (Stanford).
Ganor, B. (2001) 'Suicide Attacks in Israel', in *Countering Suicide Terrorism: An International Conference* (Herzliya, Israel).
Gates, J. M. (1990) 'People's War in Vietnam', *Journal of Military History*, 54/3, 325–44.
Gelpi, C., Feaver, P. D. and Reifler, J. (2005/6) 'Success Matters: Casualty Sensitivity and the War in Iraq', *International Security*, 30/3, 7–46.
Gilio, M. E. (1972) *The Tupamaros* (London).
Giri, S. (2008) 'Taking the Bait: Maoists and the Democratic Lure in Nepal', *Journal of Contemporary Asia*, 38/2, 277–99.
Glantz, D. M. and House, J. M. (1995) *When Titans Clashed: How the Red Army Stopped Hitler* (Kansas City).
Goodwin, J. and Skocpol, T. (1989) 'Explaining Revolutions in the Contemporary Third World', *Politics and Society*, 17/4, 489–510.
Goodwin, J. (2001) *No Other Way Out: States and Revolutionary Movements, 1945–1991* (Cambridge).
Gorman, S. M. (1982) 'The Role of the Revolutionary Armed Forces', in T. W. Walker (ed.) *Nicaragua in Revolution* (New York).
Gott, R. (1970) *Guerrilla Movements in Latin America* (London).
Gray, C. S. (2005) *Another Bloody Century: Future Warfare* (London).
Grayling, A. C. (2007) *Among the Dead Cities: Is the Targeting of Civilians in War Ever Justified?* (London).
Griffin, N. (2002) *Caucasus: Mountain Men and Holy Wars* (London).
Guevara, C. (1961) *Guerrilla Warfare* (New York and London).
Guillen, A. (1982 [1973]) 'Urban Guerrilla Strategy', in G. Chaliand (ed.) *Guerrilla Strategies* (Berkeley).
Gunaratna, R. (2002) *Inside Al Qaeda* (Carlton North, Australia).
Gunaratna, R. (2003) 'Sri Lanka: Feeding the Tamil Tigers', in K. Ballentine and J. Sherman (eds) *The Political Economy of Armed Conflict: Beyond Greed and Grievance* (Boulder).
Gurr, T. R. (1970) *Why Men Rebel* (Princeton).
Gurr, T. R. (1990) 'Terrorism in Democracies: Its Social and Political Bases', in W. Reich (ed.) *Origins of Terrorism* (Washington D. C.).
Hale, W. (1994) *Turkish Politics and the Military* (London).
Hamill, D. (1986) *Pig in the Middle: The Army in Northern Ireland 1969–1985* (London).
Hashim, A. S. (2006) *Insurgency and Counter-Insurgency in Iraq* (Ithaca, New York).
Hastings, M. (1993 [1981]) *Das Reich. The March of the 2nd SS Panzer Division Through France, June 1944* (London).

Heggoy, A. A. (1972) *Insurgency and Counterinsurgency in Algeria* (Bloomington, Indiana).
Hengel, M. (1975) *Victory Over Violence* (London).
Hess, H. (1973) *Mafia and Mafiosi: The Structure of Power* (Lexington, Mass.).
Hill, P. (2005) 'Kamikaze, 1943–5', in D. Gambetta (ed.) *Making Sense of Suicide Missions* (Oxford).
Hills, A. (2004) *Future War in Cities* (London).
Hirst, P. (2001) *War and Power in the 21st Century* (Oxford).
Hobsbawm, E. J. (1972) *Bandits* (Harmondsworth).
Hobsbawm, E. J. (1973) *Revolutionaries: Contemporary Essays* (New York).
Hodes, C. and Sedra, M. (2007) *The Search for Security in Post-Taliban Afghanistan*, Adelphi Paper 391 (London).
Hoffman, B. (1998) *Inside Terrorism* (London).
Holmes, S. (2005) 'Al-Qaeda, September 11, 2001', in D. Gambetta (ed.) *Making Sense of Suicide Missions* (Oxford).
Hopgood, S. (2005) 'Tamil Tigers, 1987–2002', in D. Gambetta (ed.) *Making Sense of Suicide Missions* (Oxford).
Horne, A. (1987) *A Savage War of Peace: Algeria 1954–1962*, Revised edition (London).
House of Commons (2006) *Report of the Official Account of the Bombings in London on 7th July 2005* (London).
Howard, M. (2002) 'What's in a Name?', *Foreign Affairs*, 81/1, 8–13.
Huntington, S. P. (1959) *The Soldier and the State* (Cambridge, Mass.).
Huntington, S. P. (1962) 'Introduction: Guerrilla Warfare in Theory and Policy', in F. M. Osanka (ed.) *Modern Guerrilla Warfare* (New York).
Huntington, S. P. (1991) *The Third Wave. Democratization in the Late Twentieth Century* (Norman, Oklahoma).
Jackson, R. H. (1990) *Quasi-States: Sovereignty, International Relations, and the Third World* (Cambridge).
Jenkins, B. M. (1986) 'Defense against Terrorism', *Political Science Quarterly*, 101/5, 773–86.
Joes, A. J. (1996) *Guerrilla Warfare* (Westport, Conn.).
Joshi, M. (1996) 'On the Razor's Edge: The Liberation Tigers of Tamil Eelam', *Studies in Conflict and Terrorism*, 19/1, 19–42.
Kalyvas, S. N. (2001) '"New" and "Old" Civil Wars: A Valid Distinction?', *World Politics*, 54/1, 99–118.
Kalyvas, S. N. (2006) *The Logic of Violence in Civil War* (Cambridge).
Kalyvas, S. N. and Sanchez-Cuenca, D. (2005) 'Killing Without Dying: The Absence of Suicide Missions', in D. Gambetta (ed.) *Making Sense of Suicide Missions* (Oxford).
Karmon, E. (2001) 'Trends in Contemporary International Terrorism', in *Countering Suicide Terrorism: An International Conference* (Herzliya, Israel).
Keating, P. (2000) 'Getting Peasants Organized', in F. Chongyi and D. Goodman (eds) *North China at War: The Social Ecology of Revolution 1937–1945* (Lanham, Maryland).
Keegan, J. (1993) *A History of Warfare* (London).
Keeley, L. H. (1996) *War Before Civilization: The Myth of the Peaceful Savage* (Oxford).
Kellman, B. (2007) *Bioviolence: Preventing Biological Terror and Crime* (Cambridge).

Kepel, G. (2002) *Jihad: The Trail of Political Islam* (Cambridge, Mass.).
Kilcullen, D. (2009) *The Accidental Guerrilla* (New York).
Kissane, B. (2005) *The Politics of the Irish Civil War* (Oxford).
Klerks, P. (2003) 'The Network Paradigm Applied to Criminal Organisations', in A. Edwards and P. Gill (eds) *Transnational Organised Crime* (London).
Kramer, M. (1990) 'The Moral Logic of Hizballah', in W. Reich (ed.) *Origins of Terrorism* (Washington D. C.).
Kramer, M. (2004/05) 'The Perils of Counterinsurgency: Russia's War in Chechnya', *International Security*, 29/3, 5–63.
Kuperman, A. J. (1999) 'The Stinger Missile and U.S. Intervention in Afghanistan', *Political Science Quarterly*, 114/2, 219–64.
Kuran, T. (1989) 'Sparks and Prairie Fires: A Theory of Unanticipated Political Revolution', *Public Choice*, 61, 41–74.
Kuran, T. (1991) 'Now Out of Never: The Element of Surprise in the East European Revolution of 1989', *World Politics*, 44/1, 7–48.
Labrousse, A. (1973) *The Tupamaros: Urban Guerrillas in Uruguay* (Harmondsworth).
Laffan, M. (2004) 'Republicanism in the Revolutionary Decade: The Triumph and Containment of Militarism', in M. J. Bric and J. Coackley (eds) *From Political Violence to Negotiated Settlement* (Dublin).
Laqueur, W. (1977) *Guerrilla: A Historical and Critical Study* (London).
Laqueur, W. (1987 [1977]) *The Age of Terrorism* (Boston).
Laqueur, W. (2001) *The New Terrorism: Fanaticism and the Arms of Mass Destruction* (London).
Lautenschlager, K. (1986–7) 'The Submarine in Naval Warfare, 1901–2001', *International Security*, 11/3, 94–140.
Lawrence, B. (2005) *Messages to the World: The Statements of Osama Bin Laden* (London).
Liberman, P. (1996) *Does Conquest Pay? The Exploitation of Occupied Industrial Societies* (Princeton).
Lohmann, S. (1994) 'The Dynamics of Informational Cascades: The Monday Demonstrations in Leipzig, East Germany, 1989–91', *World Politics*, 47/1, 42–101.
Luttwak, E. N. (1995) 'Toward Post-Heroic Warfare', *Foreign Affairs*, 74/3, 109–22.
Mack, A. (1975) 'Why Big Nations Lose Small Wars: The Politics of Asymmetric Conflict', *World Politics*, 27/2, 175–200.
Malaquias, A. (2007) 'Angola: How to Lose a Guerrilla War', in M. Boas and K. C. Dunn (eds) *African Guerrillas: Raging Against the Machine* (Boulder).
Mao Zedong (1966a [1936]) 'Problems of Strategy in China's Revolutionary War', in Mao Zedong (ed.) *Selected Military Writings of Mao Tse-tung* (Peking).
Mao Zedong (1966b [1938]) 'Problems of Strategy in Guerrilla War Against Japan', in Mao Zedong (ed.) *Selected Military Writings of Mao Tse-tung* (Peking).
Mao Zedong (1966c [1938]) 'On Protracted War', in Mao Zedong (ed.) *Selected Military Writings of Mao Tse-tung* (Peking).
Mao Zedong (1966d [1938]) 'Problems of War and Strategy', in Mao Zedong (ed.) *Selected Military Writings of Mao Tse-tung* (Peking).
Marcuse, P. (2004) 'The "War on Terrorism" and Life in Cities after September 11, 2001', in S. Graham (ed.) *Cities, War and Terrorism* (Oxford).
Marighella, C. (1971) *Minimanual of the Urban Guerrilla*. Reprinted as Appendix to R. Moss, *Urban Guerrilla Warfare* (London).

Marten, K. (2007) 'Warlordism in Comparative Perspective', *International Security*, 31/3, 41–73.
McCauley, C. (2001) 'The Psychology of Group Identification and the Power of Ethnic Nationalism', in D. Chirot and M. E. P. Seligman (eds) *Ethnopolitical Warfare: Causes, Consequences, and Possible Solutions* (Washington D. C.).
McPherson, J. M. (1996) *Drawn with the Sword: Reflections on the American Civil War* (New York).
Meijer, R. (2005) 'The "Cycle of Contention" and the Limits of Terrorism in Saudi Arabia', in P. Aarts and G. Nonneman (eds) *Saudi Arabia in the Balance* (London).
Merari, A. (1990) 'The Readiness to Kill and Die: Suicidal Terrorism in the Middle East', in W. Reich (ed.) *Origins of Terrorism* (Washington D. C.).
Merom, G. (2003) *How Democracies Lose Small Wars* (Cambridge).
Miller, J. A. (1980) 'Urban Terrorism in Uruguay: The Tupamaros', in B. E. O'Neill, W. R. Heaton and D. J. Alberts (eds) *Insurgency in the Modern World* (Boulder).
Morrison-Taw, J. and Hoffman, B. (1994) *The Urbanization of Insurgency* (Santa Monica, California).
Moss, R. (1971) *Urban Guerrilla Warfare*, Adelphi Paper No. 79 (London).
Moss, R. (1972) *Urban Guerrillas* (London).
Mueller, J. (2005) 'The Iraq Syndrome', *Foreign Affairs*, 84/6, 44–54.
Muller, E. N. and Opp, K-D. (1986) 'Rational Choice and Rebellious Action', *American Political Science Review*, 80/2, 471–88.
Murray, W. (1977) 'German Air Power and the Munich Crisis', in B. Bond and I. Roy (eds) *War and Society: A Yearbook of Military History Vol. 2* (New York).
Murray, W. (1992) 'Net Assessment in Nazi Germany in the 1930s', in W. Murray and A. R. Millet (eds) *Calculations: Net Assessment and the Coming of World War II* (New York).
Murray, W. and Millet, A. R. (1992) 'Net Assessment on the Eve of World War II', in W. Murray and A. R. Millet (eds) *Calculations: Net Assessment and the Coming of World War II* (New York).
Murray, W. and Knox, M. (2001) 'Thinking about Revolutions in Warfare', in M. Knox and W. Murray (eds) *The Dynamics of Military Revolution: 1300–2050* (Cambridge).
Nagl, J. A. (2005) *Learning to Eat Soup With a Knife: Counterinsurgency Lessons from Malaya and Vietnam* (Chicago).
9/11 Commission (2005) *Final Report of the National Commission on Terrorist Attacks Upon the United States* (New York).
Norton, A. R. (2000) 'Hizballah and the Israeli Withdrawal from Southern Lebanon', *Journal of Palestine Studies*, 30/1, 22–35.
O'Brian, P. (2003 [1988]) *The Letter of Marque* (London).
Offer, A. (1995) 'Going to War in 1914: A Matter of Honor?', *Politics and Society*, 23/2, 213–42.
Olson, M. (1971 [1965]) *The Logic of Collective Action* (Cambridge, Mass.).
O'Neill, B. E. (1990) *Insurgency and Terrorism: Inside Modern Revolutionary Warfare* (Herndon, Virginia).
O'Neill, B. E. (2005) *Insurgency and Terrorism: From Revolution to Apocalypse* (Washington D. C.).
O'Sullivan, P. and Miller, J. W. (1983) *The Geography of Warfare* (London).
Overy, R. J. (1980) *The Air War 1939–1945* (London).

Overy, R. J. (2000) *The Battle* (Harmondsworth).
Ozdag, U. and Aydinli, E. (2003) 'Winning a Low Intensity Conflict: Drawing Lessons from the Turkish Case', *The Review of International Affairs*, 2/3, 101–21.
Paget, J. (1967) *Counter-Insurgency Operations* (New York).
Pape, R. A. (2003) 'The Strategic Logic of Suicide Terrorism', *American Political Science Review*, 97/3, 343–61.
Petersen, R. D. (2001) *Resistance and Rebellion: Lessons from Eastern Europe* (Cambridge).
Pike, D. (1966) *Viet Cong: The Organization and Techniques of the National Liberation Front of South Vietnam* (Cambridge, Mass.).
Popkin, S. L. (1979) *The Rational Peasant: The Political Economy of Rural Society in Vietnam* (Berkeley).
Popkin, S. L. (1988) 'Political Entrepreneurs and Peasants Movements in Vietnam', in M. Taylor (ed.) *Rationality and Revolution* (Cambridge).
Porch, D. (2006) *Wars of Empire* (New York).
Porzecanski, A. C. (1973) *Uruguay's Tupamaros* (New York).
Putnam, R. D. (1993) *Making Democracy Work: Civic Traditions in Modern Italy* (Princeton).
Puzo, M. (1998 [1969]) *The Godfather* (London).
Ramakrishna, K. (2003) 'Making Malaya Safe for Decolonization: The Rural Chinese Factor in the Counterinsurgency Campaign', in M. Frey *et al* (eds) *The Transformation of Southeast Asia: International Perspectives on Decolonization* (New York).
Reno, W. (2000) 'Shadow States and the Political Economy of Civil Wars', in M. Berdal and D. M. Malone (eds) *Greed and Grievance: Economic Agendas in Civil Wars* (Boulder).
Ricks, T. E. (2009) *The Gamble: General David Petraeus and the American Military Adventure in Iraq, 2006–2008* (New York).
Robinson, D. H. (1980 [1962]) *The Zeppelin in Combat: A History of the German Naval Airship Division, 1912–1918* (Seattle).
Ross, M. L. (2003) 'Oil, Drugs and Diamonds: The Varying Roles of Natural Resources in Civil War', in K. Ballentine and J. Sherman (eds) *The Political Economy of Armed Conflict: Beyond Greed and Grievance* (Boulder).
Roy, O. (1991) *The Lessons of the Soviet/Afghan War* (London).
Sageman, M. (2004) *Understanding Terror Networks* (Philadelphia).
Sambanis, N. (2004) 'What is Civil War? Conceptual and Empirical Complexities of an Operational Definition', *Journal of Conflict Resolution*, 48/6, 814–58.
Schelling, T. C. (1966) *Arms and Influence* (New Haven).
Schmid, A. P. (1983) *Political Terrorism: A Research Guide to Concepts, Theories, Data Bases and Literature* (Amsterdam and New Brunswick).
Schultz, R. (1978) 'The Limits of Terrorism in Insurgency Warfare: The Case of the Viet Cong', *Polity*, 11/1, 67–91.
Scott, J. C. (1985) *Weapons of the Weak: Everyday Forms of Peasant Resistance* (New Haven).
Short, A. (1975) *The Communist Insurrection in Malaya 1948–1960* (London).
Short, P. (2004) *Pol Pot: The History of a Nightmare* (London).
Shultz, R. H. and Dew, A. J. (2006) *Insurgents, Terrorists and Militias* (New York).
Silber, L. and Little, A. (1996) *The Death of Yugoslavia*, Revised Edition (London).

Sluka, J. A. (1997) 'From Graves to Nations: Political Martyrdom and Irish Nationalism', in J. Pettigrew (ed.) *Martyrdom and Political Resistance* (Amsterdam).
Smith, M. L. R. (1995) *Fighting For Ireland? The Military Strategy of the Irish Republican Movement* (London).
Spiers, E. M. (1975) 'The Use of the Dum-Dum Bullet in Colonial Warfare', *Journal of Imperial and Commonwealth History*, 4/1, 67–91.
Sprinzak, E. (1990) 'The Psychopolitical Formation of Extreme Left Terrorism in a Democracy: The Case of the Weathermen', in W. Reich (ed.) *Origins of Terrorism* (Washington D. C.).
Stepan, A. (1998) 'Modern Multinational Democracies: Transcending a Gellnerian Oxymoron', in J. A. Hall (ed.) *The State of the Nation* (Cambridge).
Stokke, K. (2006) 'Building the Tamil Eelam State: Emerging State Institutions and Forms of Governance in LTTE-controlled areas in Sri Lanka', *Third World Quarterly*, 27/6, 1021–40.
Sweig, J. E. (2002) *Cuban Revolution: Fidel Castro and the Urban Underground* (Cambridge, Mass.).
Taber, R. (1970 [1965]) *The War of the Flea: A Study of Guerrilla Warfare, Theory and Practice*, Second edition (St. Albans, England).
Tareke, G. (2002) 'From Lash to Red Star: The Pitfalls of Counter-Insurgency in Ethiopia, 1980–82', *Journal of Modern African Studies*, 40/3, 465–98.
Tareke, G. (2004) 'From Af Abet to Shire: The Defeat and Demise of Ethiopia's "Red" Army 1988–89', *Journal of Modern African Studies*, 42/2, 239–81.
Taylor, M. (1988) 'Rationality and Revolutionary Collective Action', in M. Taylor (ed.) *Rationality and Revolution* (Cambridge).
Thomas, L. (1929) *Raiders of the Deep* (London).
Thompson, R. (1966) *Defeating Communist Insurgency* (London).
Tilly, C. (1978) *From Mobilization to Revolution* (Reading, Mass.).
Tilly, C. (1985) 'War Making and State Making as Organized Crime', in P. B. Evans, D. Rueschemeyer and T. Skocpol (eds) *Bringing the State Back In* (Cambridge).
Toft, M. D. (2003) *The Geography of Ethnic Violence* (Princeton).
Tone, J. L. (1994) *The Fatal Knot: The Guerrilla War in Navarre and the Defeat of Napoleon in Spain* (Chapel Hill and London).
Townshend, C. (1979) 'The Irish Republican Army and the Development of Guerrilla Warfare, 1916–1921', *English Historical Review*, 94, 318–45.
Townshend, C. (2002) *Terrorism: A Very Short Introduction* (Oxford).
Trotsky, L. (1934) *The History of the Russian Revolution* (London).
Tucker, R. C. (1995) *Politics as Leadership*. Revised Edition (Columbia, Missouri).
Van Creveld, M. (1991) *The Transformation of War* (New York).
Van Creveld, M. (2000) *The Art of War: War and Military Thought* (London).
Van Creveld, M. (2002) 'The Transformation of War Revisited', *Small Wars and Insurgencies*, 13/2, 3–14.
Waldmann, P. (1986) 'Guerrilla Movements in Argentina, Guatemala, Nicaragua, and Uruguay', in P. H. Merkl (ed.) *Political Violence and Terror* (Berkeley).
Wallas, G. (1924 [1908]) *Human Nature in Politics* (London).
Waltz, K. N. (1979) *Theory of International Politics* (New York).
Waterman, H. (1981)'Reasons and Reason: Collective Political Activity in Comparative and Historical Perspective', *World Politics*, 33/4, 554–89.
Weber, M. (1947 [1920]) *The Theory of Social and Economic Organization* (New York).

Weber, M. (1970 [1919]) 'Politics as a Vocation', in H. H. Gerth and C. W. Mills (eds) *From Max Weber: Essays in Sociology* (London).

Weinberg, L. (2006) 'Suicide Terrorism for Secular Causes', in A. Pedahzur (ed.) *Root Causes of Suicide Terrorism: The Globalization of Martyrdom* (London).

Weinstein, J. M. (2007) *Inside Rebellion: The Politics of Insurgent Violence* (Cambridge).

Westad, O. A. (2003) *Decisive Encounters: The Chinese Civil War, 1946–1950* (Stanford).

White, J. R. (2003) *Terrorism: An Introduction*, Fourth edition (Belmont, California).

Wickham-Crowley, T. P. (1992) *Guerrillas and Revolution in Latin America* (Princeton, New Jersey).

Windrow, M. (1997) *The Algerian War: 1954–62* (Botley, England).

Windrow, M. (1998) *The French Indochina War: 1946–54* (Botley, England).

Wood, E. J. (2003) *Insurgent Collective Action and Civil War in El Salvador* (Cambridge).

Woodiwiss, M. (2003) 'Transnational Organised Crime: The Global Reach of an American Concept', in A. Edwards and P. Gill (eds) *Transnational Organised Crime* (London).

Wright, Q. (1964 [1942]) *The Study of War*, Abridged Edition (Chicago).

Wright, T. C. (1991) *Latin America in the Era of the Cuban Revolution* (New York).

Wrong, M. (2005) *I Didn't Do It For You* (London).

Zamoyski, A. (2008) *Warsaw 1920: Lenin's Failed Conquest of Europe* (London).

Zartman, I. W. (1993) 'The Unfinished Agenda: Negotiating Internal Conflicts', in R. Licklider (ed.) *Stopping the Killing: How Civil Wars End* (New York).

Index